The Sociology of Generations

Jennie Bristow

The Sociology of Generations

New Directions and Challenges

Jennie Bristow
Department of Sociology,
Canterbury Christ Church University
Canterbury, UK

PLYMOUTH UNIVERSITY

9009582298

ISBN 978-1-137-60135-3   ISBN 978-1-137-60136-0   (eBook)
DOI 10.1057/978-1-137-60136-0

Library of Congress Control Number: 2016943275

Printed on acid-free paper

This Palgrave Macmillan imprint is published by Springer Nature
The registered company is Macmillan Publishers Ltd. London

For our kids

ACKNOWLEDGEMENTS

I have many people to thank for their role in this book. Professor Frank Furedi and Dr Ellie Lee for their contribution to the ideas presented here, and for their friendship and intellectual collaboration. Friends and colleagues at Canterbury Christ Church University and the Centre for Parenting Culture Studies, for their encouragement and inspiration. The editorial team at Palgrave Macmillan, for their interest in and enthusiasm for this project.

Toby Marshall, for sharing his insights on the new school curriculum, and Joanna Williams, for her thoughts on academic freedom and trends in higher education. David Axe, Charlotte Faircloth, Jan Macvarish, Beverley Marshall, Sally Millard, Helen Reece, Jane Sandeman, Alka Sehgal-Cuthbert, and many other friends without whom discussions of the questions raised here would be that much less fruitful or enjoyable.

My students, whose encounter with the sociology and experience of education has clarified a number of questions, and made teaching as enjoyable as it is fulfilling. And my family, for their love and support. Thank you all.

CONTENTS

CHAPTER 1

Why Study Generations?

Abstract The concept of 'generation' denotes the biological reality of being, the historical reality of living, and the epistemological problem of knowing. These multiple meanings often operate simultaneously, making generation a powerful concept for understanding the social world; and also a slippery concept, which is difficult to define and apply. This chapter summarises the ways in which sociology has approached the study of generations over the twentieth century, and, following Mannheim, situates the problem of generations within the sociology of knowledge.

Keywords Mannheim • Burke • Marx • Life course • Globalisation • Social contract

The study of generations is the study of a series of interactions, all of which occur at once. It involves relations between individual and family, between biology and society, between culture, social structures, and historical events; it is shaped by time and place, and given meaning through the context in which it occurs. No wonder the concept of generation has been redefined throughout history; no wonder its meaning remains continually contested. And little surprise that attempts to define and make sociological sense of generations often, as Philip Abrams observed, 'end up either as genealogy (the history of fathers and sons in particular families) or as waffle' (Abrams 1970, p. 176).

© The Editor(s) (if applicable) and The Author(s) 2016
J. Bristow, *The Sociology of Generations*,
DOI 10.1057/978-1-137-60136-0_1

1

Almost one hundred years ago, the sociologist Karl Mannheim sought to make sense of 'The Problem of Generations' in a way that embraced the very difficulties involved in the study of this phenomenon. The sociological significance of generations, contended Mannheim, could not be comprehended through a focus either on their quantitative existence or their qualitative experience: the sociology of generations is neither a question of numbers nor the introspective study of everyday life. What matters is the interaction between 'new participants in the cultural process' (Mannheim 1952, p. 292) and the society in which these participants are born, develop, and transform their world. In this respect, the problem of generations is the problem of knowledge: how we, as a society, ensure that the world lives on through those whom we leave behind.

Mannheim's was not the first or only attempt to theorise generations from a sociological perspective. Indeed, the first part of his essay grapples with the ideas put forward by other thinkers, from both the positivist and the romantic-historical traditions, and draws from these approaches the elements synthesised in his own formulation of the problem. Since Mannheim, there have been other important developments in the study of generations and the sociology of knowledge. These later approaches, briefly considered below, both extend and challenge Mannheim's approach, attempting to find ways of studying empirically the experience of generations, and accounting for social and cultural changes that affect the way that the problem of generations is framed and understood.

There is no scope, in this short book, to do justice to the wealth of literature that has contributed to the field over the past century, and what follows is not an attempt to synthesise all these developments. Rather, the aim is to draw out some specifically new directions and challenges that arise when we examine the problem of generations in the context of Anglo-American societies today. Generations are defined, here, neither in the narrow cohort sense (a group of people born around the same time) nor by the more individualised life course approach, but, following Mannheim, as historical, or social, generations, whose self-definition is forged by the circumstances in which they come as age. As such, we see the problem of generations as a problem of knowledge—how society's accumulated cultural heritage is transmitted from generation to generation at a time when the status both of knowledge itself, and those charged with passing it on, stands in question.

FEATURES OF THE PROBLEM OF GENERATIONS TODAY

The period between the two world wars of the twentieth century high-lighted the problem of generations as a bloody, and starkly polarised, reality in Europe. The disillusionment of the 'Generation of 1914' and the rise of the German Youth Movement, in the context of economic turmoil, cultural decadence, and intense and organised class conflict, threw into question assumptions about established truths and the enduring value of the 'old ways' (Gillis 1973; Karl 1970; Laqueur 1962; Mannheim 1952; Wohl 2009). This conflict—of politics, ideologies, and belief systems—permeated the existence and experience of those people living through the times.

Mannheim's distinctive contribution was to formulate an understanding of the emergence and operation of generational consciousness, during times of accelerated social change. In the following chapters, we discuss in more detail the process by which knowledge is transmitted, and the integration of generational location with the experience of wider social and historical events. Here, we consider some reasons why Mannheim saw the study of generations as important in the 1920s, and why this should remain the case today.

History and Biography

C. Wright Mills, writing in 1959, promulgated the sociological imagination as a way of thinking that 'enables us to grasp history and biography and the relations between the two within society'. The 'first fruit of this imagination', he wrote, 'is the idea that the individual can understand his own experience and gauge his own fate only by locating himself within his period, that he can know his own chances in life only by becoming aware of those of all individuals in his circumstances' (Mills 1970, p. 12).

For Mannheim, the importance of generations lay similarly within their temporal location—a group of people born during the same historical period and in the same geographical location, who would interact with the same social forces and events. However, he insisted, they would not experience or shape these forces in the same ways: one's generational location is only part of a broader life story, which is given meaning by other social, cultural, and familial factors. An individual's social, or class, location was the closest analogy to generational location: one does not choose the social class into which one is born, and it has a powerful effect on the way in which one knows, experiences, and shapes the world (Mannheim 1952).

In drawing comparisons with class location to elucidate his theory of generations, Mannheim did not imply that generation was the more significant. Indeed the interwar period was marked by a heightened consciousness of social class and national identity, certainly by comparison with any subsequent epochs, and these had a significant impact on individuals' identity and the mobilisation of agency. But the very features of this time that gave rise to class consciousness, and national consciousness—rapid social change, a schism between the ideas and values of the past and present, the collective shock brought about by a long, bloody, and traumatic world war—also gave rise to a distinctive *generational* consciousness.

The 'Generation of 1914', according to Wohl, comprised 'wanderers between two worlds': the traditions of the past and the uncertain present in which the young veterans of the Great War found themselves. His inspiration for this phrase derives from Aldous Huxley, writing in 1942:

> I was born wandering between two worlds, one dead, the other powerless to be born, and have made, in a curious way, the worst of both. (*Letters of Aldous Huxley*. Cited in Wohl 2009, p. 203)

Huxley, in turn, seems to recall Hamlet's anguished plaint: 'The time is out of joint. O cursèd spite, That ever I was born to set it right!' (Shakespeare 1993 [c. 1600], p. 878). The experience of time out of joint, of a schism between the past and an uncertain present, is not new. Yet to experience such a schism in terms that are self-consciously generational is generally understood to be a relatively modern phenomenon, which is linked with the development of industrial society, and with this the development of collective experience and agency. In this sense, Wohl explains, we could see generational consciousness as 'one of the side effects of the coming of mass society':

> It was, like the concept of class, a form of collectivism and determinism, but one that emphasized temporal rather than socioeconomic location. (Wohl 2009, p. 207)

The extent to which individuals make sense of their experiences in generational terms, rather than (or as well as) in terms of social class, personal identity, or political outlook, is one of the enduring debates within the sociology of generations. As I discuss in a previous study (Bristow 2015), there was a sentiment in the second half of the twentieth century that the waning of class consciousness and solidarity might result in the mobilisation of politics based on age (Abrams 1970; Goertzel 1972), as part of a general turn towards the politics of status and, later, identity. The policy discourse in the USA and Britain today is replete with claims about 'intergenerational justice', premised on the assumption that there are conflicts of interest between different age groups, which lend themselves to political action (Walker 1996; White 2013).

However, the extent to which different age groups do experience their problems in terms of a 'generation war' remains open to question and debate. Indeed, Mannheim's approach would imply that generational consciousness develops *because* of a wider sense of collective agency, rather than as a substitute for it. Below, we suggest that in recent decades the label of generation has tended to be applied from above, rather than emerging from the actions of particular, 'active', or 'strategic' (Edmunds and Turner 2002a) generations themselves.

Continuity and Change

Marx's famous passage on the making of history, published in 1852, emphasised the extent to which the agency of the present is constrained and shaped by the past:

> Man makes his own history, but he does not make it out of the whole cloth; he does not make it out of conditions chosen by himself, but out of such as he finds close at hand. The tradition of all past generations weighs like an alp on the brain of the living. At the very time when men appear engaged in revolutionizing things and themselves, in bringing about what never was before, at such very epochs of revolutionary crisis do they anxiously conjure up into their service the spirits of the past, assume their names, their battle cries, their costumes to enact a new historic scene in such time-honored disguise and with such borrowed language. (Marx 2011 [1852], p. 1)

As the times move on, and the present is created, the tradition of the past is assimilated and transcended. 'Thus does the beginner, who has acquired a new language, keep on translating it back into his own mother tongue; only then has he grasped the spirit of the new language and is able freely to express himself therewith when he moves in it without recollections of the old, and has forgotten in its use his own hereditary tongue,' argued Marx (2011, p. 1).

Burke, on the other side of the revolutionary divide, saw the French Revolution in terms of a crisis of generations, the effect of which could only be destructive. In a passage often cited by conservative thinkers in the present day, he described society as a 'contract' to be looked upon with reverence

> because it is not a partnership in things subservient only to the gross ani-mal existence of a temporary and perishable nature. It is a partnership in all science, a partnership in all art, a partnership in every virtue and in all perfection. As the ends of such a partnership cannot be obtained in many generations, it becomes a partnership not only between those who are living, but between those who are living, those who are dead, and those who are to be born. (Burke 2014 [1790], loc. 1429)

For Burke, and his followers in the present day, the contract between generations is what provides stability between the past, present, and future. For Marx, Mannheim, and others who conceive a more dynamic approach to history and history-making, the importance of generations lies in their relation to the process of change. New generations are not conceived of, or celebrated as, a complete break from the past—what is special about the younger generation, from this perspective, is that it comes to the world with both a connection to the past and an orientation to the present.

As the sociology of generations developed in the post-war period, the literature has tended to explain theories in terms of two opposing 'camps': the structural-functionalist perspective, associated with Talcott Parsons (1963) and S. N. Eisenstadt (1956, 1963, 1971); and the 'historical-consciousness' perspective associated with Mannheim (1952). The implication is that the structural-functionalist approach is, like Burke, mainly concerned with social stability and continuity with the past, whereas Mannheim's approach is one that embraces change and is concerned with generations in their relation to the future. Such phrases as '[t]he tradition of all past generations weighs like an alp on the brain of the living' (Marx 2011, p. 1) are interpreted one-sidedly, as a call-to-arms for the younger generation to throw off the shackles of its past.

However, the polarisation of the structural-functional and the historical-consciousness approaches does not do justice to the subtlety with which each engages with the question of generational continuity and change. The tendency to politicise generation theory, presenting one as 'conservative' and the other as 'revolutionary', fails to appreciate that the core of all serious studies of the problem of generations relates to the problem of knowledge: how a society derives meaning from its past in order to shape the future.

Thus Eisenstadt's seminal study of age groups and social structure, *From Generation to Generation* (1956), and the contributions made by Eisenstadt and Parsons to Erikson's (1963) study of youth in America following the Second World War, share with Mannheim a sensitive appreciation of the potentially fraught relationship between traditional social norms and values and the expectations of a younger generation growing up during a period where such norms have been recently shattered. We explore this point further in Chap. 3.

In the present day, the US writers Strauss and Howe, authors of several books including the hubristic *Generations: The History of America's Future, 1584 to 2069* (1991) and *The Fourth Turning: An American Prophecy – What the Cycles of History Tell Us About America's Next Rendezvous with Destiny* (1998), seek to identify historical patterns or cycles that could predict which generations would become more or less influential. They are influenced in this endeavour by the work of Ortega y Gasset in the 1930s, whose 'pulse-rate hypothesis' of generational change, which sought 'the regularities of the universal rhythm of generations' (Jaegar 1985, p. 280), is generally recognised by scholars as extreme, even 'outlandish', because of 'its imposition of biological rhythms on socio-historical phenomena' (Dobson 1989, p. 176). Yet the influence of Strauss and Howe's work today reveals an ongoing search for naturalistic theories and a yearning that the sociology of generations should provide us with a blueprint for the future.

The recent policy focus on generations has often (over)emphasised demographic shifts, leading to a preoccupation with absolute or relative cohort size that obscures other important factors in generational influence or experience: for example, changing economic circumstances, or cultural trends (Howker and Malik 2010; Willetts 2010). As we discuss in Chap. 5, explanations of generational conflict based on population numbers have tended towards determinism and one-sided conclusions (see discussion in Bristow 2015). However, demography at its best can provide a powerful account of generation as a sociological category, within the broader context in which it exists (Davis 1940; Lesthaeghe 2010; MacInnes and Díaz 2009).

In this essay, we emphasise that the aim of a historical approach to generations is not to seek quasi-biological rhythms to explain generational identity but to understand the mediations between biology, culture, family, and society, within specific historical moments. Mannheim presented the 'problem of generations' as a dynamic interaction between cohorts of individuals, the tempo of wider social change, and cultural moments (the *Zeitgeist*). In both the interwar and post-war periods, dislocation and disorientation provided the basis for an emergent generational consciousness—a distinctive interpretation of the *Zeitgeist*, born out of the experience of coming of age at a time that is, more than usually, out of joint.

Gender, Reproduction, and Life Course

The 'Generation of 1914', writes Wohl, was 'a self-image produced by a clearly-defined group within the educated classes', which 'derived its credibility and its force from circumstances that were unique to European men born during the last decades of the nineteenth century' (Wohl 2009, p. 209). Indeed, when Mannheim discussed the problem of generations, his focus was on European men. Women were not, at that point, explicitly part of the story of generations.

The story of generations today looks quite different. Women in the early twenty-first century are not only part of this story, but increasingly, its authors. The implications of this for the sociology of generations are complex. The centrality of women to the modern story of generations has helped to foster a wealth of insightful literature that explores the experience of generations over the life course. Following the pioneering work of Hareven (1978, 1991, 1994, 2000) and others, life course approaches complicate pre-existing categories of 'age' or 'life stages', recognising that definitions of generations, and interactions between generations, are more fluid and shaded than is often assumed.

Life course sociology has variously challenged assumptions about the meaning of youth, ageing, and personal life (Phillipson 2013; Smart 2007). It has contributed a rich seam to the sociology of family life, underpinning discussions about parenting, fathering, the meaning of time, and the experience of adult intergenerational relationships (Brannen 2006, 2015; Treas et al. 2014). In this regard, life course sociology provides an invaluable approach to empirical study, particularly within a context where age and family norms have lost much of their clarity (Pilcher 1995). The qualitative dimension of these approaches has brought a subtlety to research about such ephemeral aspects of personal existence as love, commitment, and intimacy that has previously been under-acknowledged (Carter 2013; Faircloth 2015), and contributes to and complicates influential sociological theories about the 'transformation of intimacy' (Giddens 1992). The emergence of birth control technologies in the 1920s, and their normalisation in the 1960s and 1970s, provided women with both the ability to control their fertility and the expectation that they would do so (Beck-Gernsheim 2002; Lupton 2013; Marks 2010). The institutionalisation of 'family planning' norms, alongside structural changes to women's position in the labour market and cultural shifts in assumptions about women's position, has brought changes both to the *biological* story of generations and the cultural norms attached to reproduction.

In this regard, the greater inclusivity of the study of generations has also brought some tensions to the fore. As women have become part of the social story of generations, the functions and relations associated with reproduction and the private realm have become increasingly rationalised, and subject to bureaucratic norms and rules. These trends are discussed in Chaps. 4 and 5, which indicate some of the ways in which an explicit social policy focus on the problem of generations assumes a brittleness about generational interaction, which intersects with and fuels cultural trends towards individuation and fragmentation.

Boundaries and Labels

The impact of globalisation also raises some new questions. Mannheim (1952) emphasised the importance of geographical location in the emergence of generational consciousness, remarking that, for example, 'young people in Prussia about 1800 did not share a common generation location with young people in China at the same period' (Mannheim 1952, p. 303). In experiential terms, the social and cultural changes experienced in different countries and cultures continue to shape the development of generational consciousness in heterogeneous ways. But we might also expect that the sense of generational identity that is forged at a time where there is not only greater connectedness between cultures, via mass travel and the media, but also a far greater sensibility of this connectedness, has a more diffuse character.

The growth of the internet and television media has contributed to a certain 'de-nationalisation' of popular culture and communication within Anglo-American culture, leading some to challenge the importance of geographical boundaries in the formation of generational experience and identity (Edmunds and Turner 2005). This cultural process has mirrored the wider trend towards globalisation, the fragmentation of national boundaries within Europe, and the promotion of 'global education' (Standish 2012).

'It has become clear that shifts in the global economy, the transnational dissemination of ideas, and new forms of biopolitics are reconfiguring the nature of childhood, youth and old age,' write Cole and Durham (2007). However, they continue:

> But children, youth and the elderly do not reconceptualize themselves or conform to new social patterns as isolated individuals or age groups. Rather, it is in the relationships between age groups that changes take shape, as people negotiate pragmatically and emotionally to manage the present and to reproduce desirable and livable futures. (Cole and Durham 2007, pp. 2–3)

Cole and Durham's edited collection, *Generations and Globalization*, draws together case studies from diverse countries—China, Mexico, Botswana, India—thereby combining an appreciation of national trends and contexts in relation to wider globalising cultural and economic trends. Likewise, Edmunds and Turner's edited volume offers a valuable discussion of 'Generational Consciousness, Narrative, and Politics' (2002b) through essays focusing on women, national struggles, ethnic identities, and age groups.

The 'globalisation' of generational identity is not an entirely recent phenomenon. If the 'Generation of 1914' was the story of European men, the 'Baby Boomer' generation of men and women is an identity understood in several national contexts, including Australia, France, Britain, and North America. This was a generation forged by its relation to a global war and the social, cultural, and economic upheavals that followed it, which we now term, simply, 'The Sixties'. Indeed, Edmunds and Turner (2005) argue that 'the 1960s generation was the first global generation, the emergence of which had world-wide consequences; today with major developments in new electronic communications, there is even more potential for the emergence of global generations that can communicate across national boundaries and through time' (Edmunds and Turner 2005, p. 559). In looking at the cultural construction of the Boomer generation, we can see that many of the themes and motifs have come about as a result not of indigenous features of a cohort within a particular country but as a result of the 'cross-national diffusion' (Best 2001) of ideas about what this generation was seen to represent (see discussion in Bristow 2015). The role of culture in the formation and labelling of generations has become better understood in recent years, drawing on the work of Becker, Bourdieu, and other cultural sociologists (Eyerman and Turner 1998; Edmunds and Turner 2002a, b). Cultural expressions of generational identity and conflict have historically been significant, as evidenced through their expression in literature (for example, Turgenev's classic 1862 novel *Fathers and Sons*, or the writings of the First World War poets and the Beats). Generational identity and agency is best understood not as a purely cultural trend, but rather a mediation between culture, social, personal, and historical factors.

We might note, for example, that the label popularly ascribed to the generation that followed the Boomers is 'Generation X', based on a novel by the Canadian writer Douglas Coupland (1991), which caught something of the *Zeitgeist* of the generation that came of age in the 1990s. This label has proved more powerful and enduring in the British context than its alternative, 'Thatcher's Children', which related more specifically to the British 'Generation X' cohort, who grew up in a political landscape dominated by the Conservative prime minister Margaret Thatcher between 1979 and 1990 (Pilcher and Wagg 2005).

As we examine the generations coming of age in the more recent past, what is most striking is the way in which labels tend to be searched for and applied in advance of—or in place of—generational self-definition. The label of 'Generation X' stuck, in part, because it chimed with the now-notorious cultural and political passivity of the 'lost' generation born in the shadow of the Boomers. The lack of imagination in the labels generally bestowed upon subsequent generations—Generation Y, Millennials, and now apparently Generation Z—reveals a cultural search for generational distinctiveness in the absence of any apparent agency. The trundle towards the end of the alphabet conveys a wider sensibility of the end of history. With characteristic wit and perspicacity, one of Coupland's recent novels is titled *Generation A* and imagines a world in the near future

where bees are extinct, until five unconnected people around the world—in the US, Canada, France, New Zealand, and Sri Lanka—are all stung. Their shared experience unites them in ways they never could have imagined. (Coupland 2010, back cover)

One paradox, therefore, of the heightened focus on generation in the present day is that it represents a search for a new narrative, rather than a response to a distinctly new generational consciousness. Generations are seen to be brought together by accident (a bee sting) rather than action. Generational labels tend to be applied globally, often with little regard for cultural and experiential differences, and politicised by claims-makers—as in *The Jilted Generation*, a term created by the journalists Ed Howker and Shiv Malik (2010) to further their arguments for 'intergenerational justice'. This floundering quest indicates that the crisis at the heart of the contemporary generation debate is a crisis of knowledge.

GENERATIONS AND THE SOCIOLOGY OF KNOWLEDGE

In formulating 'the sociological problem of generations', Mannheim concerned himself with two main, and related, questions. One was the development of generations themselves. How does a particular section of society become a generation, developing a sense of generational consciousness, and having a particular effect upon the *Zeitgeist*? The second was the question of intergenerational relations. How is the 'accumulated cultural heritage' transmitted from generation to generation, and how is knowledge transformed in the process?

A century on, both of these questions are as pertinent, and contested, as they were in the interwar years. Mannheim's complaint was that attempts to understand the problem at that time provided 'a striking illustration of the anarchy in the social and cultural sciences, where everyone starts out afresh from his own point of view … never pausing to consider the various aspects as part of a single general problem, so that the contributions of the various disciplines to the collective solution could be planned' (Mannheim 1952, p. 287).

While he recognised that starting out from a particular viewpoint was to some extent 'both necessary and fruitful', Mannheim's aim was to position sociology as 'the organising centre' for work on the problem by other disciplines. This required a sociological transition, 'from the formal static to the formal dynamic and from thence to applied historical sociology—all three together comprising the complete field of sociological research' (Mannheim 1952, pp. 287–8).

The study of generations, then, requires the study of man's social existence, the study of social dynamism, and the study of history. To understand the process of generational transmission requires a consideration of natural and cultural factors, the wider social forces operating in a particular time and place, the experience of history, and personal development. Given this, it is not surprising that Mannheim's theory is difficult to apply to the empirical study of generational experience or relations (Pilcher 1994). By a similar token, there is no suggestion here that Mannheim could, or should, be 'used' in place of contributions from other branches of sociology or disciplines within the social sciences or humanities, which provide distinct insights into the workings of generational consciousness, relations between the generations, and the cultural or political meanings attached to genera-tion in the past and modern world.

Rather, we can see Mannheim's approach as a guide to the problem of generations in its totality: the realm of experience that is covered, and the variety of social and individual factors that are touched upon, when we talk about 'a generation'. As we consider generations within their wider his-tory, we can begin to understand the import of certain temporal, cultural, and social changes on the ways that emerging generations come to make sense of their world—and conversely, the ways that the world makes sense of the young people coming through.

In the remainder of this essay, we focus on two significant changes that should be read into Mannheim's understanding of the problem of generations, as it reveals itself in the present day. The first is to do with the way that society's accumulated cultural heritage is transmitted to younger generations, and the particular difficulties that arise from anxieties about the norms, values, and knowledge of the past. This pertains both to the project of formal education and to the project of socialisation—the twin processes of 'social remembering' and 'enabling us to forget' (Mannheim 1952, p. 294). Here, we suggest that an ongoing discomfort with estab-lished knowledge and the 'old ways' has the effect of disturbing both the 'consciously recognized models on which men patterned their behaviour', and the 'unconsciously "condensed", merely "implicit" or "virtual" pat-terns' that Mannheim identified as equally important in the process of generational renewal (Mannheim 1952, p. 295).

The second change relates to the interplay between natural and social factors, and how this gives rise to a particular generational consciousness. Specifically, we examine the intersection between the technologies of fertility control, and social changes with regard to women's roles and responsibilities, in the context of a wider consciousness of individuation and risk. We discuss how 'generation' has become an explicit focus for policy discourse, which has the effect (and perhaps the intention) of bringing the relationships of the private realm under the purview of bureaucratic management. The relationship between biology and culture here is not a straightforward one and needs to be considered within its broader context, of ambivalence about knowledge and history and a bureaucratisation of relations between, and within, generations. However, we should recognise that the change in the role of women and the ways in which society has come to conceptualise the norms of reproduction in both its biological and social senses represent a distinct feature of the problem of generations in the present day, in ways that could not have been predicted a century ago.

This essay takes Mannheim as its starting point and intellectual guide, but owes an equal debt to the work of those who have theorised risk society, education, and intimate life in the later twentieth century, in particular Frank Furedi, Christopher Lasch, Anthony Giddens, Elisabeth Beck-Gernsheim, and Michael Young. It is informed by historical research conducted by the author on the history of children's reading and the history of the British Abortion Act, both of which reveal how society's anxiety about the problem of generations runs through ongoing debates about knowledge, education, and the regulation of reproduction.

While the essay seeks to engage with this 'problem of generations', it does not seek to answer every question, or to resolve the problem itself. As Mannheim so eloquently argued, the tensions that arise from society's relationship with its younger generations are precisely what give this process its creativity and dynamism, and allow us to keep renewing and shaping our world.

REFERENCES

Abrams, P. (1970). Rites de passage: The conflict of generations in industrial society. *Journal of Contemporary History*, 5(1) Generations in Conflict, 175–190.

Beck-Gernsheim, E. (2002). *Reinventing the family: In search of new lifestyles.* Cambridge: Polity Press.

Best, J. (Ed.). (2001). *How claims spread: Cross-national diffusion of social problems.* New York: Aldine de Gruyter.

Brannen, J. (2006). Cultures of intergenerational transmission in four-generation families. *The Sociological Review*, 54(1), 133–154.

Brannen, J. (2015). *Fathers and sons: Generations, families and migration.* Basingstoke: Palgrave Macmillan.

Bristow, J. (2015). *Baby boomers and generational conflict*. Basingstoke: Palgrave Macmillan.

Burke, E. (2014 [1790]). *Reflections on the revolution in France*. (e-book) SMK Books.

Carter, J. (2013). The curious absence of love stories in women's talk. *The Sociological Review, 61*, 728–744.

Cole, J., & Durham, D. (2007). *Generations and globalization: Youth, age and family in the new world economy*. Bloomington/Indianapolis: Indiana University Press.

Coupland, D. (1991). *Generation X: Tales for an accelerated culture*. London: Abacus Books.

Coupland, D. (2010). *Generation A*. London: Windmill Books.

Davis, K. (1940). The sociology of parent-youth conflict. *American Sociological Review, 5*(4), 523–535.

Dobson, A. (1989). *An introduction to the politics and philosophy of José Ortega y Gasset*. Cambridge: Cambridge University Press.

Edmunds, J., & Turner, B. S. (2002a). *Generations, culture and society*. Buckingham/Philadelphia: Open University Press.

Edmunds, J., & Turner, B. S. (Eds.). (2002b). *Generational consciousness, narrative and politics*. London/Boulder/New York/Oxford: Rowman & Littlefield Publishers, Inc.

Edmunds, J., & Turner, B. S. (2005). Global generations: Social change in the twentieth century. *The British Journal of Sociology, 56*(4), 559–577.

Eisenstadt, S. N. (1956). *From generation to generation: Age groups and social structure*. New York/London: The Free Press.

Eisenstadt, S. N. (1963). Archetypal patterns of youth. In E. H. Erikson (Ed.), *Youth: Change and challenge*. New York/London: Basic Books.

Eisenstadt, S. N. (1971). Generational conflict and intellectual antinomianism. *Annals of the American Academy of Political and Social Science, 395* Students Protest, 68–79.

Erikson, E. H. (Ed.). (1963). *Youth: Change and challenge*. New York/London: Basic Books.

Eyerman, R., & Turner, B. S. (1998). Outline of a theory of generations. *European Journal of Social Theory, 1*(1), 91–106.

Faircloth, C. (2015). Negotiating intimacy, equality and sexuality in the transition to parenthood. *Sociological Research Online, 20*(4), 3.

Giddens, A. (1992). *The transformation of intimacy: Love, sexuality and eroticism in modern societies*. Cambridge: Polity Press.

Gillis, J. R. (1973). Conformity and rebellion: Contrasting styles of English and German youth, 1900–33. *History of Education Quarterly, 13*(3), 249–260.

Goertzel, T. (1972). Generational conflict and social change. *Youth and Society, 3*(3), 327–352.

Hareven, T. K. (1978). The search for generational memory: Tribal rites in industrial society. *Daedalus 107*(4), Generations, 137–149.

Hareven, T. K. (1991). The history of the family and the complexity of social change. *The American Historical Review, 96*(1), 95–124.

Hareven, T. K. (1994). Aging and generational relations: A historical and life course perspective. *Annual Review of Sociology, 20*, 437–461.

Hareven, T. K. (2000). *Families, history, and social change: Life-course and cross-cultural perspectives.* Boulder/Oxford: Westview Press.

Howker, E., & Malik, S. (2010). *Jilted generation: How Britain has bankrupted its youth.* London: Icon.

Jaeger, H. (1985). Generations in history: Reflections on a controversial concept. *History and Theory, 24*(3), 273–292.

Karl, W. (1970). Students and the youth movement in Germany: Attempt at a structural comparison. *Journal of Contemporary History, 5*(1) – Generations in Conflict, 113–127.

Laqueur, W. Z. (1962). *Young Germany: A history of the German Youth Movement.* London: Routledge/Kegan Paul.

Lesthaeghe, R. (2010). The unfolding story of the second demographic transition. *Population and Development Review, 36*(2), 211–251.

Lupton, D. (2013). *The social worlds of the unborn.* Basingstoke: Palgrave Macmillan.

MacInnes, J., & Díaz, J. P. (2009). The reproductive revolution. *The Sociological Review, 57*(2), 262–284.

Mannheim, K. (1952). *Essays on the sociology of knowledge.* Paul Kecskemeti (Ed.). London: Routledge & Kegan Paul Ltd.

Marks, L. V. (2010). *Sexual chemistry: A history of the contraceptive pill.* New Haven: Yale University Press.

Marx, K. (2011 [1852]). *The eighteenth Brumaire of Louis Bonaparte.* (e-book) CreateSpace Independent Publishing.

Mills, C. W. (1970). *The sociological imagination.* London: Penguin.

Parsons, T. (1963). Youth in the context of American society. In E. H. Erikson (Ed.), *Youth: Change and challenge.* New York/London: Basic Books.

Phillipson, C. (2013). *Ageing.* Cambridge: Polity Press.

Pilcher, J. (1994). Mannheim's sociology of generations: An undervalued legacy. *British Journal of Sociology, 45*(3), 481–495.

Pilcher, J. (1995). *Age and generation in modern Britain.* Oxford: Oxford University Press.

Pilcher, J., & Wagg, S. (2005). *Thatcher's children? Politics, childhood and society in the 1980s and 1990s.* London/New York: Routledge.

Shakespeare, W. (1993 [c. 1600]). Hamlet (Act One, Scene Five). In: *The complete works of William Shakespeare.* London: Robinson Publishing.

Smart, C. (2007). *Personal life: New directions in sociological thinking.* Cambridge: Polity Press.

Standish, A. (2012). *The false promise of global learning: Why education needs boundaries.* London/New York: Bloomsbury.

Strauss, W., & Howe, N. (1991). *Generations: The history of America's future, 1584 to 2069.* New York/London: Harper Perennial.

Strauss, W., & Howe, N. (1998). *The fourth turning: An American prophecy – What the cycles of history tell us about America's next rendezvous with destiny.* New York: Broadway Books.

Treas, J., Scott, J., & Richards, M. (2014). *The Wiley Blackwell companion to the sociology of families.* West Sussex: John Wiley & Sons.

Walker, A. (1996). *The new generational contract: Intergenerational relations, old age and welfare.* London: UCL Press.

White, J. (2013). Thinking generations. *British Journal of Sociology, 64*(2), 216–247.

Willetts, D. (2010). *The pinch: How the baby boomers took their children's future— And why they should give it back.* London: Atlantic Books.

Wohl, R. (2009). *The generation of 1914.* Cambridge: Harvard University Press.

CHAPTER 2

Fresh Contacts, Education, and the Cultural Heritage

Abstract Education is properly understood as a generational responsibility, in which the accumulated cultural heritage is passed on to students who, because they have grown up in different times, will take and shape this knowledge in their own way. This chapter develops Mannheim's understanding of the importance of 'fresh contacts' to discuss the crisis of the curriculum over the twentieth century, where ambivalence about the cultural heritage has allowed instrumental imperatives to dominate the purpose of education.

Keywords Knowledge • School • Hannah Arendt • Curriculum • Michael Gove • Michael Young

The past, wrote L.P. Hartley in 1953, 'is a foreign country: they do things differently there' (Hartley 1997, p. 5). But in the early twenty-first century, the past is positioned not just as foreign but alien; bereft of guides about what we should be passing on to our children, and replete with cautionary tales of what not to think, believe, or do. No sooner, it seems, are younger generations assimilated into the cultural heritage than they are incited to transcend it. The intellectual impact of post-modernism and its associated trends has been charged with fragmenting the very notion of a cultural heritage; the ascendancy of identity politics demands that individuals construct their own, personalised past(s) (Bloom 1987; Williams 2016).

J. Bristow, *The Sociology of Generations*,
DOI 10.1057/978-1-137-60136-0_2

Formal education over the course of the twentieth century has placed increasing emphasis on instrumental goals, such as the development of particular attitudes or employment skills, and less on historical understanding and traditional knowledge (Oakeshott 2001; Young 2008; Young and Lambert 2014). Ongoing disputes over curriculum content and the role of the teacher have positioned as problematic both the cultural heritage and those charged with transmitting it to the younger generation (Ball 2013; Gewirtz and Cribb 2009; Ward and Eden 2009). Debates about education are characterised by a tension between a stated desire to 'pass on the past' and a continual corruption of this goal by the imperative of preparing younger generations for the 'real world' of today (Finn 2015; Furedi 2009).

Mannheim's account of the transmission of 'the accumulated cultural heritage' (Mannheim 1952, p. 292) discussed the role played by both formal education and the passing on of 'virtual' or unconscious data through formative experiences and intergenerational contact. At both these levels, the generational transmission of cultural heritage has come to be perceived as problematic. Informal encounters within the family, and between adults and children in the community at large, have become a focus for anxiety about the promotion of outdated attitudes or processes of socialisation; and the socialising role of the family is increasingly complemented, or challenged by, a more explicit socialising function played by the school and other agencies.

Chap. 4 examines how the formalisation of intergenerational encounters, expressed by the imperative of 'safeguarding', raises new questions about the relation between adults and children, as the embodiment of the past-present and the present-future, respectively. The location of older generations within the past and the present underpins their relationship of authority to the young: the wisdom of their experience is what enables them to perceive the harms and opportunities of the world around them, so as to help young people navigate this world and develop their own future. We discuss how this relationship between the generations has increasingly become mediated, indeed distanced, by cultural anxieties and bureaucratic mechanisms that seek to shield young people from the experience, and authority, of the adult world.

This chapter and the next focus on the project of education: how, as a society, we work through the mechanisms by which we transmit the 'accumulated cultural heritage' to our young. Mannheim saw the role of education—'[t]he data transmitted by conscious teaching'—as having a 'more limited importance' than implicit knowledge gained through 'the automatic passing on to the new generations of the traditional ways of life, feelings, and attitudes' (Mannheim 1952, p. 299). However, it is through debates about 'conscious teaching' that our present-day anxiety about the cultural heritage, and the relation of adults to children, is most clearly revealed.

Although the focus of this essay is largely on policy developments within Britain—the trajectory of these developments and the cultural dynamics that help to drive them—readers will recognise many similar features within the USA. This reflects the extent to which certain key trends are common to much of the Western world, and certainly Anglo-American societies. 'The general crisis that has overtaken the modern world everywhere and in almost every sphere of life manifests itself differently in each country, involving different areas and taking on different forms,' wrote Hannah Arendt in the early 1960s. She argued that the politicised 'crisis in education' in post-war America revealed a far greater, and wider, anxiety about relations between the generations and 'between past and future' (Arendt 2006, p. 170). Arendt argued further that:

[T]here is always a temptation to believe that we are dealing with specific problems confined within historical and national boundaries and of importance only to those immediately affected. It is precisely this belief that in our time has consistently proved false. One can take it as a general rule in this century that whatever is possible in one country may in the foreseeable future be equally possible in almost any other. (Arendt 2006, p. 171)

In order to understand why the crisis in education has become such an enduring preoccupation within Anglo-American culture, we begin by reviewing Mannheim's insights into the process by which the transmission of the accumulated cultural heritage from the older to the younger generation occurs, and why this is vital in the development of consciousness. Mannheim, and subsequently Arendt, placed the problem of generations within the framework of social and cultural *renewal*. As such, we argue that the purpose of education should not be to shape young people in the image of the past, nor to equip them with the skills to navigate the demands of the present, but to give them the foundations to work out their own future, according to the circumstances in which they find themselves.

FRESH CONTACTS AND THE CULTURAL HERITAGE

To appreciate 'which features of social life result from the existence of generations', Mannheim suggests that we 'make the experiment of imagining what the social life of man would be like if one generation lived on for ever and none followed to replace it' (Mannheim 1952, p. 292). Against this 'utopian, imaginary society', he works out the 'basic phenomena implied by the mere fact of the existence of generations' in our own society:

(a) new participants in the cultural process are emerging, whilst
(b) former participants in that process are continually disappearing;
(c) members of any one generation can participate only in a temporally limited section of the historical process, and
(d) it is therefore necessary continually to transmit the accumulated cultural heritage;
(e) the transition from generation to generation is a continuous process. (Mannheim 1952, pp. 292–3)

In this way, Mannheim places the biological realities of birth and death at the centre of the problem of generations. Because nobody lives forever, and because there is indeed one born every minute, there can be no regimented or tidy way of ensuring the maintenance of society's accumulated cultural heritage. Rather, this has to be transmitted continuously, 'from generation to generation', via a combination of 'conscious teaching' and informal mechanisms of generational interaction. As he explains:

> [A] utopian, immortal society would not have to face this necessity of cultural transmission, the most important aspect of which is the automatic passing on to the new generations of the traditional ways of life, feelings, and attitudes. The data transmitted by conscious teaching are of more limited importance, both quantitatively and qualitatively. All those attitudes and ideas which go on functioning satisfactorily in the new situation and serve as the basic inventory of group life are unconsciously and unwittingly handed on and transmitted: they seep in without either the teacher or pupil knowing anything about it. (Mannheim 1952, p. 299)

The fact that we do not live in a 'utopian, immortal society' is, for Mannheim, the source of knowledge's dynamism. It means that our culture is never merely preserved, but that it is constantly developed by 'fresh contacts' with the accumulated heritage; a process that 'always means a changed relationship of distance from the object and a novel approach in assimilating, using, and developing the proffered material' (Mannheim 1952, p. 293).

A similar process is expressed in Arendt's discussion of natality. Arendt, like Mannheim, understood generations as having both a social and a natural existence, and education as an important way in which both elements of this existence were mediated. Our central concern with regard to education, she argued, is 'the relation between grown-ups and children in general or, putting it in even more general and exact terms, our attitude toward the fact of natality: the fact that we have all come into the world by being born and that this world is constantly renewed through birth.' Arendt continued:

> Education is the point at which we decide whether we love the world enough to assume responsibility for it and by the same token save it from that ruin which, except for renewal, except for the coming of the new and the young, would be inevitable. And education, too, is where we decide whether we love our children enough not to expel them from our world and leave them to their own devices, nor to strike from their hands their chance of undertaking something new, something unforeseen by us, but to prepare them in advance for the task of renewing a common world. (Arendt 2006, p. 193)

We can understand by the term 'education' both the formal education of children in schools and the ways in which children are socialised into prevailing social norms—by families, schools, youth groups, and other adults within the community. The changes identified here with regard to both education and socialisation derive from a wider crisis of knowledge: and viewed historically, these processes should be understood as interdependent. Certainly their consequences for the transmission of accumulated cultural heritage from generation to generation follow the same direction. That is, where Mannheim saw the locus for the dynamic construction of knowledge within the younger generation and their fresh contacts with all that was known before, the dynamic in the present day is to foreshorten this process, through an ambivalence and uncertainty about, and sometimes outright disdain for, the cultural heritage at the point at which it is passed on.

TENSIONS WITHIN, AND BEYOND, THE CURRICULUM

In their discussion of curriculum developments in British education, Moore and Young (2001) present two models. The first is 'neo-conservative traditionalism', which perceives the curriculum as 'a given body of knowledge that it is the responsibility of the schools to transmit'. This model, claim Moore and Young, 'is as old as the institution of schooling itself', and implies both the study of particular works—for example, the canon of English literature—and a particular relationship: a 'relationship of deference to a given body of knowledge', which is 'inspired by the view that the traditional discipline of learning promotes proper respect for authority and protects traditional values' (Moore and Young 2001, p. 447).

The second model is 'technical-instrumentalism'. For proponents of the technical-instrumental model, 'the curriculum imperative is not educational in the traditional sense, but supportive of what they see as the needs of the economy'; indeed, '[f]rom this perspective, education, the curriculum and even knowledge itself becomes a means to an end, not an end in itself' (Moore and Young 2001, p. 447). The tension between these two models, argue Moore and Young, 'has influenced the development of the curriculum for more than a century' (p. 448). Indeed, this is a tension, not only over the content of the curriculum, but over the wider meaning of education and how the problem of generations is mediated.

To put it baldly: proponents of the technical-instrumental perspective see the generational transmission of the cultural heritage as a by-product of the proper function of education, which is conceived as giving children the skills to participate in the workforce. From this perspective, the importance of social remembering is subservient to the immediate imperative that young people should be taught how to meet the demands of 'now'.

Proponents of the conservative traditionalist perspective, meanwhile, tend to conceive the transmission of the cultural heritage as a rigid, one-way process, which can be added on to and on top of other, instrumentalising tendencies within the management of education. Thus, teachers are instructed to apply 'traditional' forms of knowledge according to present-day methods of monitoring and accountability, and what becomes an attempt to transmit the cultural heritage can end up further distancing the younger generation from its past.

Below, we discuss some of the ways in which these tensions over the content of the curriculum—and, by extension, the meaning of education—have revealed themselves over the past century. At every turn, it is vital to look at these developments in their social and historical context: ideas about generations, and the relations between them, are shaped by the wider social forces of their times, and do not develop in one clear direction. However, the common trajectory of education policy in recent decades has been one that fundamentally questions the idea that the primary purpose of education should be the transmission of the accumulated cultural heritage, and that the teacher is best placed to mediate this generational transaction.

INSTRUMENTALISM VERSUS HUMANISM: A LONG-RUNNING TENSION

Debates about the project of education—and specifically, mass education provided through schools—have for a long time been framed by wider interests and social, political, cultural, and economic concerns. This can be clearly seen in Britain during the interwar years: a time of palpable crisis, when education was already becoming invested with the power to ameliorate a range of problems, from children's physical ill-health to the nation's improved economic performance.

For example, in a speech to the House of Lords on 12 July 1916, Viscount Haldane, former Secretary of State for War and Lord Chancellor, 'called attention to the training of nation and the necessity of preparing for the future', arguing that 'The task is to prepare the future generation—morally, physically, and intellectually—to endure the strain.' But as Lord Haldane emphasised the need for more extensive educational training of the young in the interests of economic competitiveness, 'Earl Cromer said he could conceive no greater disaster than to put the whole education of the country on a utilitarian basis':

> The moral collapse of Germany was one of the most extraordinary and tragic events in history. Side by side with a great advance in material prosperity and scientific attainments was a great deterioration of the German character. One of the causes was the atmosphere created by too little attention being given to humanistic and classical literature. (*Manchester Guardian* 1916)

The First World War, explains Mathieson (1975), 'uncovered the old problems of the elementary schools—large classes, poorly qualified staff, physically weak children, children in part-time employment, and continued use of outdated, discredited methods of mechanical rote-learning'. One of the war's results was 'to produce not only a sense of the military and economic benefits enjoyed by Germany because of her educational system's freedom from irrelevant traditionalism but also an awareness of our working-class's cultural inferiority' (Mathieson 1975, pp. 69–70). Thus, Lloyd George proclaimed, in 1918:

> The most formidable institution we had to fight in Germany was not the arsenals of Krupps or the yards in which they turned out submarines, but the schools of Germany. They were our most formidable competitors in business and our most terrible opponents in war. An educated man is a better worker, a more formidable warrior, and a better citizen. That was only half comprehended before the war. (Cited in Mathieson 1975, p. 70)

The war also focused attention on education as a solution to Britain's domestic problems. H.A.L. Fisher, speaking in 1917 in preparation for his Bill to raise the school-leaving age, said:

> I conceive that it is part of the duty of our generation to provide some means for compensating the tragic loss which our nation is enduring, and that one means by which some compensation may be provided is by the creation of a system of education throughout the country which will increase the value of every human unit in the whole of society by giving all our children the best possible opportunity that we can afford to give them, and they can afford to turn to account. (Cited in Mathieson 1975, p. 70)

At this time, the generational responsibility of education was conceptualised in terms of economic necessity, social stability, and moral regeneration. However, the question of how to define and transmit the cultural heritage was already emerging as contradictory and fraught, with instrumental imperatives clashing with an uneasy promotion of traditional values.

One example of the way this played out is given by the debate about the teaching of English as a subject of schools. The Newbolt Report, published in 1921 and formally titled *The Teaching of English in England*, provides a clear example of the tensions that came to the fore during this period. The Newbolt Report—named after the poet Sir Henry Newbolt, chair of the Departmental Committee appointed to 'inquire into the position of English in the educational system of England'—aimed to distil the position of English with regard to three main questions: '(1) the requirements of a liberal education; (2) the needs of business, the professions, and public services; and (3) the relation of English to other studies' (Newbolt Report, p. 1).

The vision enshrined in the Newbolt Report draws heavily on Matthew Arnold's (2015 [1869]) approach to the question of 'Culture and

Anarchy', which views culture as a panacea for a wide range of social and moral problems. The Report claimed that education should be 'divided into the training of the will (morals), the training of the intellect (science) and the training of the emotions (expression or creative art)' (Newbolt Report, p. 9). While science was framed as 'the methodical pursuit of truth and the conquest of the physical world by human intelligence and skill', literature was held to have a more spiritual quality, encompassing the generational transmission of experience and worthy qualities:

> Literature, the form of art most readily available, must be handled from the first as the most direct and lasting communication of experience by man to men. It must never be thought of or represented as an ornament, an excrescence, a mere pastime or an accomplishment; above all, it must never be treated as a field of mental exercise remote from ordinary life. The sphere of morals in school life is limited by practical considerations with which we cannot here deal, but it is evident that if science and literature can be ably and enthusiastically taught, the child's natural love of goodness will be strongly encouraged and great progress may be made in the strengthening of the will. The vast importance to a nation of moral training would alone make it imperative that education shall be regarded as experience and shall be kept in the closest contact with life and personal relations. (Newbolt Report 1921, p. 9)

Mathieson explains that both the Newbolt Report and George Sampson's (1970 [1922]) book, *English for the English*, with which the Report shares many ideas and language, 'express all the major anxieties' about the treatment of English Literature in universities, schools, and teacher-training establishments at the time, 'as well as all the certainties about the value of English which had been intensified since Arnold's analysis of his "mechanical" and "external" society. They reflect, too, the characteristic mood of the period following the First World War, the sharp despair and the faith in the power of education to improve the future' (Mathieson 1975, p. 69).

The authors of the Newbolt Report self-consciously asserted the case for the study of literature in moral, humanistic, and spiritual terms. They rejected outright the suggestion that the teaching of English should pay greater heed to 'the needs of business', recommending that '"the needs of business" are best met by a liberal education' and that the promotion of ' "Commercial English" is objectionable to all who have the purity of the language at heart, and also unnecessary' (Newbolt Report 1921, p. 351).

The report also contains a powerful anti-industrial sensibility, commonly articulated by Victorian 'preachers of culture' (Mathieson 1975; Williams 1971). As such, it reads as much as a yearning for a lost world of community and certainty as a self-confident attempt to forge a competitive, classless society where, as Arnold would have it, culture 'seeks to do away with classes; to make the best that has been thought and known in the world current everywhere; to make all men live in an atmosphere of sweetness and light' (Arnold 2015, p. 44, loc 871).

Yet there is an instrumental dimension to the Newbolt Report, which betrays the extent of anxiety about the process of transmitting the cultural heritage to the younger generation. 'Expressions of national guilt and the need for greater social justice were now being made publicly at an official level, and it was becoming clear that recommendations for the replacement of the classics by English studies were having implications far beyond practical changes in the curriculum,' argues Mathieson (1975, p. 71). The Newbolt Report's call for a move away from the study of Classics, which to that point had formed the core of the public school curriculum, was motivated both by the alleged inaccessibility of the Latin language and Classical culture to a wider mass of children, and by the power invested in English literature to promote a sense of national identity and citizenship. This, in turn, was seen to compensate for the apparent spiritual crisis of the times. Literature, the Report argued, 'is not just a subject for academic study, but one of the chief temples of the human spirit, in which all should worship' (Newbolt Report, p. 259).

Pike's (2006) analysis of 'the secularization of literacy and the moral education of citizens' confirms the extent to which the reading of literature has long been endowed with spiritual qualities. Mathieson explains how that '[t]he arts' embodiment of spiritual values, a process hastened during the Victorian period by artists' alienation from their society, meant that they came to be recommended with greater and greater fervour for the majority's well-being.' For example, Sampson refers to the 'class of young barbarians whose souls are to be touched by literature' and to the 'pure religion' and 'creative reception' of the literary experience (Mathieson 1975, p. 76).

The promotion of English literature as a panacea for society's present-day ills endowed the cultural heritage with a powerful moralising mission. 'For fifty years teachers have been trying to make the elementary schoolboy *know* something, when they should have been trying to make him *be* something,' proclaimed Sampson. 'They have been trying to make him, not a man, but an epitome of information' (Sampson 1970, p. 36; emphasis in original). The sense that the project of education was about forming a particular kind of character—making the schoolboy '*be* something'—was bolstered by the wider cultural trappings of upper-class adolescence in the Victorian era. It was character embodied by Rudyard Kipling's (1895) poem *If—*, in which character is learned from the wisdom of the older generation (Kipling 2001, p. 605). It is also the basis of Sir Henry Newbolt's most famous poem, 'Vitaï Lampada' (Newbolt 1892), which presents the spirit gained through playing cricket matches on the fields of England's public schools as forging the character necessary for battle.

The Boy Scouts was founded in England by Robert Baden-Powell in 1908, a movement that, explains the historian John Gillis, 'was properly Victorian in its morality and staunchly patriotic in its politics'. Gillis situates Scouting within the 'whole child saving movement of the turn of the century', and the instability of the period:

> Worried by reports of physical and moral deterioration that followed the Boer War, and anxious about the spread of secularism and socialism, the English elites moved to insulate the young of all classes, but particularly the lower orders, against the real and imagined dangers of the pre-World War I period. (Gillis 1973, p. 252)

Jenny Holt's fascinating study shows the themes promoted by boys' school stories, which, up until the First World War, similarly aimed to popularise Victorian themes about 'ideal' character and citizenship (Holt 2008, p. 209).

Yet by 1921, the character type idealised by Kipling, Newbolt, Baden-Powell, and other 'staunch' Victorians was already looking like a relic of the imagination of a previous era. 'During and after the First World War the entire thrust of the school genre changed,' writes Holt, of the literature aimed at schoolboys. 'Indeed, from the postwar period onwards, the political and pedagogical confidence of writers was so shaken that it is often hard to identify any coherent message at all' (Holt 2008, p. 209). The same was true for the wider literary field. While the Newbolt Report sought to promote a cultural heritage clearly bounded in traditional norms and an imperial, English spirituality and self-identity, this heritage was already being questioned. Thus, 'where Sampson sees literature evangelically as a source of spiritual revelation and portrays English teachers as the savers of lost uncultured souls, the Report usually prefers the language of institutional religion and of the church establishment, seeing literature as a kind of liturgical text, creating social harmony through familiar verbal patterns,' writes Scott (1990, pp. 227–8). He continues:

> Sampson was insisting literary culture gave access to unitary spiritual values, while the committee Report looked to literary culture only for a harmony of surface signs, verbal defences, of a kind that, like the established Church of England and its Common Prayer, or the reading aloud of the King James Version for its sentence patterns, allowed an almost infinite multiplicity in ideas about what exactly was being signified. (Scott 1990, pp. 227–8)

POST-WAR CURRICULUM CONFLICTS

From the vantage point of today, the Newbolt Report reads as a strikingly self-confident, humanistic, and jingoistic statement of the pre-eminence of the British Empire. There is little doubt that Sir Henry Newbolt himself brought a missionary zeal to his promotion of English, as a civilising mission both at home and in the wider Empire. In one letter to his wife, dated 2 February 1927, he writes of spending 'a morning at Paternoster Row and Whitehall Gds. and an afternoon at the Central Council for care of Churches ... and tea at the Colonial Office with Hans Vischer, who seems glad to have me on the Education Committee for making Niggers into Noble Natives on the principles of the Newbolt Report'—a proposition that is 'of course quite in the direct line for me and I'm looking forward to it' (Newbolt 1942, p. 350).

Yet by the 1970s, the assumptions enshrined in the Newbolt Report—about the content of the literary canon, the role of liberal education, and the humanising possibilities afforded by access to culture—were widely contested across Britain, the USA, and Europe. The publication of influential texts such as *Schooling in Capitalist America* (Bowles and Gintis 1976), *Learning to Labour* (Willis 1977), *Learning to Lose* (Spender and Sarah 1988 [1980]), and *Reproduction in Education, Society and Culture* (Bourdieu and Passeron 2000 [1977]) revealed the extent to which core assumptions about the cultural heritage—what it was, how it was transmitted, and why it was important—were being challenged as mechanisms that reproduced inequalities of class, race, and gender (Gewirtz and Cribb 2009).

'Increasingly, critical theorists have turned their attention from literary interpretation itself to the social and educational institutions through which literary ideas have been generated and mediated; to the young turks of the new cultural studies, the [Newbolt] Report has presented a very easy target,' writes Scott (1990). Noting that '[t]he committee's eponymous chairman, Sir Henry Newbolt himself, is remembered almost solely as an 1890s imperialist balladeeer', Scott explains:

> To the cultural analysts, little further research must have seemed necessary before Newbolt's Report could be reinterpreted to prove that literary study was an upper-class conspiracy, a false substitute for critical and class consciousness. 'It is no accident,' asserts Terry Eagleton, that 'the most influential Government report' on English teaching was written by 'a minor jingoist poet.' (Scott 1990, p. 222)

The reduction of the Newbolt Report to its imperialist assumptions and moralising mission reflects a wider turn in the sociology of knowledge, in which attempts to theorise the ways in which knowledge is constructed and transmitted became incorporated into theories that saw the promotion of the cultural heritage, and the institutions of education, as mechanisms by which a powerful (rich, white, Establishment) elite maintained its domination (Williams 2016).

These criticisms were not without foundation. As we have discussed, even by the 1920s both the content of the curriculum and the presumed purpose of schooling were shaped by powerful instrumental concerns that variously plundered or dismissed aspects of the accumulated cultural heritage, according to present-day needs and concerns. As we see below and in the next chapter, the instrumentalist imperative has overshadowed education policies over the twentieth century. In this regard, it is right to acknowledge the degree to which the institutions of education, such as schools and, in recent years, universities, construct younger generations politically and economically, as future citizens or workers, and generally serve to reproduce rather than resolve deep-seated social conflicts and inequalities (Ball 2013).

It is also important to acknowledge the extent to which the school curriculum has become a flashpoint for disputes that are at least as political as they are educational. As Ward and Eden note, of the introduction of the National Curriculum in Britain in the 1980s:

> To get a nation of 60 million people to agree on what should count as knowledge was going to be a tall order, and it took some four years

from the conception of the curriculum to its implementation. It produced a
remarkable tale of argument, intrigue and manipulation—battles between
government and its civil servants, professionals and academics. (Ward and
Eden 2009, p. 69)

Debates about 'what should count as knowledge' and therefore what
should be taught in the classroom have been inextricably bound up with
attempts to bring teachers into line with the economic and political objec-
tives of the present day. Although by comparison with many other countries, explain Ward and Eden, 'Britain was a late starter' in developing a
National Curriculum, with the passage of the 1988 Education Act by the
Conservative government, 'it made up by creating probably the world's
most detailed and rigorous national curriculum, and one that was to be
assessed by nationally standardised tests: a pincer movement on the professionals' (Ward and Eden 2009, p. 69). Centralised control over the curriculum went alongside other changes enshrined in the 1988 Act, including
'local management of schools (LMS) making head teachers into business
managers, as against their previous role as senior teacher' (Ward and Eden
2009, p. 74).

In this way, attempts to affirm society's responsibility to its younger
generations by promoting the importance of transmitting the accumulated cultural heritage have gone alongside reforms that undermine teachers' authority, autonomy, and professional role. The result is that the
contradiction between education as a generational responsibility and the
political and instrumental imperatives that have come to inform schooling
has become increasingly stark.

GOVE VERSUS THE BLOB

If there were a modern incarnation of Sir Henry Newbolt, it could be represented by the figure of Michael Gove, who served as Secretary of State
for Education in the Conservative-Liberal Democrat Coalition government
from 2010 to 2014. Gove, formerly assistant editor of the *Times* (London),
was an energetic and, from the start, controversial Education Secretary,
who would go on to usher in a number of significant reforms to the curriculum, assessment, organisation, and funding of state education.

The foundations for what would quickly be identified as 'the Gove
legacy' (Finn 2015) were laid in a speech delivered to the Royal Society
of Arts (RSA) in 2009 titled, 'What is education for?' This was an interest-

ing and important speech for a number of reasons. Gove began by complaining that, under the previous New Labour government, there was no 'single department of state charged with encouraging learning, supporting teaching and valuing education'. Instead, the New Labour government had created the Department for Children, Schools and Families (DCSF), which saw schools 'as instruments to advance central government's social agenda', and the Department for Business, Innovation and Skills (BIS), which promoted universities 'as instruments to advance central government's economic agenda' (Gove 2009, p. 2).

'What we do not have—and what we desperately need—is a Department at the heart of Government championing the cause of education, the value of liberal learning, the wider spread of knowledge as an uncontested good in its own right,' argued Gove (2009, p. 2). This echoed the Newbolt Report's call for 'a liberal education for all English children whatever their position or occupation in life' (Newbolt Report 1921, p. 14); and indeed, many of the themes in the speech speak to Arnold's (1869) vision of culture as a means to human perfectibility and social cohesion.

For Gove, the central purpose of education could be summed up as 'the democratic intellect—every citizen's right to draw on our stock of intellectual capital.' Education 'is a good in itself—one of the central hallmarks of a civilized society—indeed the means by which societies ensure that everything which is best in our society is passed on to succeeding generations' (Gove 2009, p. 2). He went on to cite Michael Oakeshott's argument that 'every human being is born heir to an inheritance—"an inheritance of human achievements; an inheritance of thoughts, beliefs, ideas, understandings, intellectual and practical enterprises, languages, canons, works of arts, books musical compositions and so on"', and argued:

> Education should be a process of granting every individual their rights to that inheritance. Every child should have the chance to be introduced to the best that has been thought, and written. To deny children the opportunity to extend their knowledge so they can appreciate, enjoy, and become familiar with the best of our civilization is to perpetuate a very specific, and tragic, sort of deprivation. (Gove 2009, p. 3)

In attacking the 'quite indefensible assumption among some that the only cultural experiences to which the young are entitled, or even open, are those which have a direct, and contemporary, relevance to their lives' (Gove 2009, p. 3), Gove criticised the instrumentalism that has informed education policy in recent decades, and was especially apparent in policies promoted by the previous, New Labour government (Moore and Young 2001). In place of an education that prioritised life skills and job skills, and reduced subjects such as English and Mathematics to the functional skills of 'literacy' and 'numeracy', there was a suggestion that the younger generation should be provided with a solid grounding in historical knowledge, the scientific method, and the canon of English literature.

From the start, Gove's self-conscious traditionalism was controversial. One flashpoint came in 2014, during a row over US writer John Steinbeck's classic novel *Of Mice and Men*. Gove, it was widely reported, had 'banned' this novel—along with the works of Arthur Miller and Harper Lee— from the school curriculum. In the ensuing furore, Gove was accused of a backward-looking philistinism, plucking a curriculum 'straight out of the 1940s' and dictating works of literature that today's pupils would find 'tedious'. Christopher Bigsby, professor of American Studies at the University of East Anglia, wrote a shrill tirade in the *Guardian* complaining that 'the union jack of culture' was 'fluttering from education central':

> As the home secretary does her best to patrol our borders to keep out international students, whom she regards as immigrants, so the GCSE syllabus is to be kept for the English for fear that Romanian novels might move in next door. (in Kennedy 2014)

Gove had not, in fact, 'banned' American books from the school curriculum. As he explained in a robust riposte in the *Daily Telegraph*, his intention was to insist that children read more literature, not less (Gove 2014). *Of Mice and Men* is a great work of modern literature: but schools seemed to be attracted to it mainly because it was short, and could be made 'relevant' to children's lives through reading themes such as bullying into the novel. The row that ensued between Gove and his critics revealed, in part, longstanding tensions and disputes about the role of literature in the school curriculum, discussed above in relation to the 1921 Newbolt Report. But more starkly, it brought to the fore the bitter tensions between Gove and the new educational establishment.

Gove had pushed his educational reforms forward in the face of opposition by teachers' unions and others, whom he termed 'the Blob', after the 1958 film starring Steve McQueen in which a 'slimy, ruthless, voracious' amoeba alien 'consumes everything in its path' (*Guardian* 2013). In reports of Gove's battle with 'the Blob', we seem to be presented with two opposed forces. One is the imperative of traditional, liberal education, which emphasises the need for children to study an academic curriculum; the other consists of those who see the institutions and ideas of education as being about preparing children for the world as it is today, and explicitly promotes schools as sites for social engineering. 'Owing to the awesome relentlessness of the Blob, nothing in the battle between traditionalists and progressives in education ever gets definitively settled in the traditionalists' favour,' wrote Dennis Sewell (2010) in the *Spectator* magazine.

As Sewell notes, 'the Blob' is not a new phenomenon, or even an original—or distinctly British—insult. The metaphor was first used to depict the educational establishment in the mid-1980s, when it was 'adopted by William Bennett, education secretary in the Reagan administration, as a term to describe the amorphous coalition of a bloated education bureaucracy, teacher unions and education research establishment that Bennett argued always obstructs or stifles school reform.' In Britain today, according to Sewell:

> The Blob currently has the whole schools system firmly in its grip. From Whitehall it issues diktats: the Children's Plan, Every Child Matters, instructions on personalised learning, safeguarding guidelines, frameworks and so forth. The Children's Services departments of local authorities provide a second tier of bureaucratic meddling while skimming off cash badly needed by schools. Through the Training and Development Agency and the National College of School Leadership, the Blob indoctrinates young teachers and determines both their teaching methods and their professional development. It has hijacked the National Curriculum and rewritten it, taken control of the Specialist Schools and Academies Trust, and has even subsumed Ofsted as callously as in the movie it swallowed a janitor. (Sewell 2010)

In declaring his determination to wrestle control of education from the bureaucratic processes and political agendas of 'the Blob', Gove sought to reverse what he saw as 'the drift from "education, education, education" to "everything else matters"'. 'I worry that our schools are being asked to do more and more which, while it might appear desirable, dilutes the importance of teaching and learning,' he argued in 2009. 'I fear that duties on schools, and teachers, to fulfil a variety of noble purposes—everything from promoting community cohesion to developing relationships with other public bodies, trusts, committees and panels gets in the way of their core purpose—education' (Gove 2009, p. 5).

The ever-expanding policy remit given to education has been the subject of much debate and critique (see Ball 2013; Chitty 2014; Tomlinson 2005). As we discuss below and in the next chapter, the trend by which schools have been required 'to fulfil a variety of noble purposes' to the detriment of their focus on the transmission of knowledge has developed over several decades, and became the dominant rhetoric for reform under the New Labour government (1997–2010). And indeed, it is significant that despite Gove's rhetoric in 2009, this trajectory towards 'reconstructing the nature of educational problems and redistributing blame' (Ball 2013, p. 179) would continue with the Conservative-Liberal Democrat Coalition government. The 'new kinds of policy solutions and methods of policy' that followed this approach, explains Ball, are ' "joined up" in two senses':

> First, solutions to educational problems are sought in part through changes in forms of governance. Second, educational problems are linked both with the needs of the economy and to social problems, for example, through 'failing' parents and 'dysfunctional families' to disaffection, truancy, school and social exclusion and crime and anti-social behaviour. (Ball 2013, p. 179)

The 'awesome relentlessness of the Blob' seems to derive less from the voraciousness of progressive educators and the bureaucratic instruments that they control than from the ever-present chinks in the traditionalists' armour of liberal education. Over the course of the twentieth century, those seeking to defend the central purpose of education as the transmission of the existing cultural heritage from the older generations to the younger have compromised their arguments with instrumental or political agendas.

THE MARKET, THE STATE, AND THE JUGGERNAUT OF INSTRUMENTALISM

While Gove cited Oakeshott's vision of education as a cultural legacy, in practice the mission of current reforms shows a further preoccupation with what Oakeshott (2001) describes as goals 'extrinsic' to education. Even while emphasising the importance of a liberal education, in practice the trajectory of recent education policy has followed and extended the instrumentalist mission of previous decades, positioning education as 'a magic bullet—the "escalator" for social mobility, a vital engine of "human capital" formation through the development of skills for the economy' (Finn 2015). It is in these terms, explains Finn, that:

[T]he contemporary political economy of education in Britain constructs the priorities of the educational agenda, elucidated in the life of the coalition government through a vocabulary of 'competitiveness' in a 'global race'. (Finn 2015, p. 2)

Gove's critics tend to present such developments as evidence of a 'neo-liberal' approach to education, which seeks to transform the project of education from a 'public good' into a marketised commodity, for the benefit of individuals. This criticism is borne out by recent changes to the funding and management of schools, which have pursued the development of 'academies' based on the principles of corporate funding and accountability to parents, and universities, where the introduction of tuition fees has clearly positioned students in a consumer role to their own, personal 'university experience' (Evans 2004; Williams 2010). In schools and universities alike, the obsession with students gaining the requisite grades and skills to be 'employable' in an internationally competitive market underwrites parental demands, inspection criteria, and the ethos—if not the content—of education.

In this respect many aspects of the current education system mimic developments in the USA. As in the USA, regimes of constant testing, monitoring, and measuring, and a growing emphasis on behaviourism as the dominant form of pupil management, can be challenged for their narrowly instrumental focus. The ongoing debate about the No Child Left Behind Act encapsulates many of these concerns. The 2002 US legislation, argue Petersen and West (2003), 'redirects educational thinking along new channels' of management and accountability:

Under its terms, every state, to receive federal aid, must put into place a set of standards together with a detailed testing plan designed to make sure the standards are being met. Students at schools that fail to measure up may leave for other schools in the same district, and, if a school persistently fails to make adequate progress toward full proficiency, it becomes subject to corrective action. (Petersen and West 2003, pp. 1–2)

While the intentions of this legislation—ensuring that 'no child' is 'left behind'—may be laudable, its consequences have been widely criticised as perverse. Adam Urbanski, vice president of the American Federation, writes in his Foreword to Hayes's (2008) critical account of *No Child Left Behind: Past, Present, and Future*:

NCLB attaches high stakes to standardized tests, narrows the curriculum, labels schools unfairly, siphons away much-needed funds from impoverished districts and schools, and allows privateers to prey on public-school children. More and more teachers tell me, 'I love to teach, but I hate my job'. (Hayes 2008, p. viii)

Critics of the 'marketisation' of education in its various forms tend to focus on the explicitly corporate rhetoric and mechanisms that now dominate policy endeavours and debates. 'Increasingly, the vocabulary of a market-based ideology substitutes the discourse of self-reliance and competition for the language of democratic participation, community, and the public good,' argues the US education professor and cultural critic Henry Giroux, in his discussion of 'The Abandoned Generation'. 'One striking example can be seen in the corporate language of schooling, in which the rhetoric of competition, self-reliance, and individual choice dominate the discourse of high-stakes testing, the standards movement, the school choice agenda, and the charter school movement' (Giroux 2003, p. 33).

But while these trends are problematic, it is important to acknowledge the extent to which instrumentalist agendas in education have also been pursued and institutionalised via mechanisms of public funding, political involvement, and state management. Because policymakers have placed an increasing value on the role of education in solving myriad problems of social (in)justice, state-funded schools and institutions of higher education have for several years found themselves organised around principles extrinsic to the passing on of knowledge.

In British policy, the orientation of education towards the needs of employers, and the positioning of pupils and their parents as consumers of education, began with a speech by the Labour Prime Minister James (Jim) Callaghan, given to Ruskin College, Oxford, in 1976. This speech, which is widely regarded as having begun the 'Great Debate' about the nature and purpose of public education (Gillard 2011; Ward and Eden 2009), is a clear statement of the government's intention to use education for instrumental concerns.

In 1973, Labour's Secretary of State for Education Anthony Crosland famously complained that the school curriculum was 'a secret garden in which only teachers and children are allowed to walk' (Ward and Eden 2009, p. 68). Callaghan's 1976 speech followed this theme by asserting the role of politicians in determining what should be taught in schools. 'It is almost as though some people would wish that the subject matter and purpose of education should not have public attention focused on it: nor that profane hands should be allowed to touch it,' stated Callaghan. He continued:

> I cannot believe that this is a considered reaction. The Labour movement has always cherished education: free education, comprehensive education, adult education. Education for life. There is nothing wrong with non-educationalists, even a prime minister, talking about it again. Everyone is allowed to put his oar in on how to overcome our economic problems, how to put the balance of payments right, how to secure more exports and so on and so on. Very important too. But I venture to say not as important in the long run as preparing future generations for life. R. H. Tawney, from whom I derived a great deal of my thinking years ago, wrote that the endowment of our children is the most precious of the natural resources of this community. So I do not hesitate to discuss how these endowments should be nurtured. (Callaghan 1976)

The assertion that the role of education should be 'preparing future generations for life' makes an important assumption about the role of education in the transmission of cultural heritage. The emphasis is firmly placed on the needs of the present, and the content of the school curriculum presented as something to be determined, not by teachers as custodians of the accumulated cultural heritage, but by a wider society—framed here as industry and 'the public'.

These sentiments were echoed, and developed, twenty years later in a speech given by Tony Blair, leader of the New Labour party, again at Ruskin College. Blair saw education as a panacea for the problems facing society in the present day: to the extent that he famously declared that his 'three priorities for government would be education, education and education'. But unlike Arnold, the promise of education was not given by access to culture—'the best which has been thought and said in the world' (Arnold 2015, p. 7, loc. 49)—but by its ability to endow the younger generation with skills and attitudes deemed appropriate for today.

Blair argued that 'our economic success and our social cohesion' depend on the success of the ability of Britain's education system to meet a number of instrumental goals, beginning with focusing on basic standards in literacy and numeracy; he emphasised that this was a project involving many social actors other than teachers:

> We will expect education—and other public services—to be held accountable for their performance; we will urge teachers to work in partnership with parents, business and the community; and we will balance parents' rights with a recognition of their responsibilities. (Blair 1996)

For the New Labour government, the need for education to serve a 'practical' purpose was so obvious and overwhelming it should put an end to all discussion. 'I believe there is the chance to forge a new consensus on education policy,' he stated. 'It will be practical not ideological. And it will put behind us the political and ideological debates that have dominated the last thirty years' (Blair 1996).

CONCLUSION

Tony Blair's determination to end the 'Great Debate' that his predecessor began reveals the extent of the turn against knowledge that characterised education in Britain at the end of the twentieth century. In the next chapter, we discuss the intellectual currents that underpinned this policy approach, and the way it reframed teaching, less as a generational responsibility, than as a technical skill.

In this context, Gove's determination to open up the question of 'what is education for?' provided a welcome recognition that debates about the transmission of the accumulated cultural heritage remain crucial to society's understanding of generational responsibility, and the ways in which this is enacted. Unfortunately, this narrative shares with previous governments a disparaging mistrust of the teaching profession, with the result that meaningful interactions between the generations continue to be compromised.

REFERENCES

Arendt, H. (2006 [1961]). *Between past and future: Eight exercises in political thought*. London: Penguin Books.

Arnold, M. (2015 [1869]). *Culture and anarchy*. [e-book] New York: Open Road Integrated Media.

Ball, S. J. (2013). *The education debate* (2nd ed.). Bristol: The Policy Press.

Blair, T. (1996, December 16). Speech given at Ruskin College, Oxford. Available at: http://www.leeds.ac.uk/educol/documents/000000084.htm. Accessed 7 Dec 2015.

Bloom, A. (1987). *The closing of the American mind: How higher education has failed democracy and impoverished the souls of today's students*. New York: Simon and Schuster.

Bourdieu, P., & Passeron, J.-C. (2000 [1977]). *Reproduction in education, society and culture*. Thousand Oaks: Sage.

Bowles, S., & Gintis, H. (1976). *Schooling in capitalist America: Educational reform and the contradictions of economic life*. New York: Basic Books.

Callaghan, J. (1976, December 18). A rational debate based on the facts. Ruskin College Oxford. Reproduced at: http://www.educationengland.org.uk/documents/speeches/1976uskin.html. Accessed 18 Dec 2015.

Chitty, C. (2014). *Education policy in Britain* (3rd ed.). Basingstoke: Palgrave Macmillan.

Evans, M. (2004). *Killing thinking: The death of universities*. London: Continuum.

Finn, M. (2015). *The Gove legacy: Education in Britain after the coalition*. Basingstoke: Palgrave Macmillan.

Furedi, F. (2009). *Wasted: Why education isn't educating*. London/New York: Continuum.

Gewirtz, S., & Cribb, A. (2009). *Understanding education: A sociological perspective*. Cambridge: Polity Press.

Gillard, D. (2011). *Education in England: A brief history*. Available at: www.educationengland.org.uk/history. Accessed 9 Dec 2015.

Gillis, J. R. (1973). Conformity and rebellion: Contrasting styles of English and German youth, 1900–33. *History of Education Quarterly, 13*(3), 249–260.

Giroux, H. A. (2003). *The abandoned generation: Democracy beyond the culture of fear*. Basingstoke: Palgrave Macmillan.

Gove, M. (2009, June 30). What is education for? Speech to the RSA. Available at: https://www.thersa.org/globalassets/pdfs/blogs/gove-speech-to-rsa.pdf. Accessed 8 Dec 2015.

Gove, M. (2014, May 26). Kill a Mockingbird? I'd never dream of it. *Daily Telegraph*. Available at: http://www.telegraph.co.uk/education/educationopinion/10857133/Michael-Gove-Kill-a-Mockingbird-Id-never-dream-of-it.html. Accessed 18 Dec 2015.

Guardian. (2013, October 2). Why does Michael Gove keep referring to the Blob?. Available at: http://www.theguardian.com/politics/shortcuts/2013/oct/02/michael-gove-referring-to-the-blob. Accessed 18 Dec 2015.

Hartley, L. P. (1997 [1953]). *The go-between*. London: Penguin.

Hayes, W. (2008). *No Child Left Behind: Past, present, and future*. Maryland: Rowman & Littlefield Education.

Holt, J. (2008). *Public school literature, civic education and the politics of male adolescence*. Surrey: Ashgate.

Kennedy, M. (2014, May 25). To Kill a Mockingbird and Of Mice and Men axed as Gove orders more Brit lit. *Guardian*. Available at: http://www.theguardian.com/education/2014/may/25/mockingbird-mice-and-men-axed-michael-gove-gcse. Accessed 7 Dec 2015.

Kipling, R. (2001). *Collected poems of Rudyard Kipling*. Hertfordshire: Wordsworth Editions.

Manchester Guardian. (1916, July 13). The training of the nation: Lord Haldane and the after-war struggle.

Mannheim, K. (1952). *Essays on the sociology of knowledge*. Paul Kecskemeti (Ed.). London: Routledge & Kegan Paul Ltd.

Mathieson, M. (1975). *The preachers of culture: A study of English and its teachers.* London: George Allen and Unwin Ltd.

Moore, R., & Young, M. (2001). Knowledge and the curriculum in the sociology of education: Towards a reconceptualisation. *British Journal of Sociology of Education, 22*(4), 445–461.

Newbolt, H. (1892). Vitaï Lampada. Available at: http://www.poemhunter.com/poem/vita-lampada/. Accessed 5 Jan 2015. Accessed 9 Dec 2015.

Newbolt, M. (Ed.). (1942). *The later life and letters of Sir Henry Newbolt.* London: Faber and Faber Ltd.

Newbolt Report. (1921). *The teaching of English in England.* London: HMSO. Available at: http://www.educationengland.org.uk/documents/newbolt/newbolt1921.html. Accessed 9 Dec 2015.

Oakeshott, M. (2001). *The voice of liberal learning.* Indianapolis: Liberty Fund.

Peterson, P. E., & West, M. R. (Eds.). (2003). *No Child Left Behind? The politics and practice of school accountability.* Washington, DC: Brookings Institution Press.

Pike, M. (2006). From beliefs to skills: The secularization of literacy and the moral education of citizens. *Journal of Beliefs & Values: Studies in Religion & Education, 27*(3), 281–289.

Sampson, G. (1970 [1921]). *English for the English: A chapter on national education.* Cambridge: Cambridge University Press.

Scott, P. (1990). English studies and the cultural construction of nationality: The Newbolt Report reexamined. In P. Scott & P. Fletcher (Eds.), *Culture and education in Victorian England* (pp. 218–232). Lewisburg/London/Cranbury: Bucknell University Press/Associated University Presses.

Sewell, D. (2010, January 13). Michael Gove vs the Blob. *Spectator.* Available at: http://new.spectator.co.uk/2010/01/michael-gove-vs-the-blob/. Accessed 8 Dec 2015.

Spender, D., & Sarah, E. (Eds.). (1988 [1980]). *Learning to lose: Sexism in education.* London: The Women's Press.

Tomlinson, S. (2005). *Education in a post welfare society* (2nd ed.). Buckingham: Open University Press.

Ward, S., & Eden, C. (2009). *Key issues in education policy.* Thousand Oaks: Sage.

Williams, R. (1971). *Culture and society 1780–1950.* London: Penguin.

Williams, J. (2010). *Consuming higher education: Why learning can't be bought.* London: Continuum.

Williams, J. (2016). *Academic freedom in an age of conformity: Confronting the fear of knowledge.* Basingstoke: Palgrave Macmillan.

Willis, P. (1977). *Learning to labour: How working class kids get working class jobs.* Farnborough: Saxon House.

Young, M. (2008). *Bringing knowledge back in: From social constructivism to social realism in the sociology of education.* Oxford/New York: Routledge.

Young, M., & Lambert, D. (2014). *Knowledge and the future school: Curriculum and social justice.* London: Bloomsbury.

CHAPTER 3

Teachers, the End of Ideology, and the Pace of Change

Abstract Teachers, as representatives of the older generation, are charged with responsibility for transmitting the cultural heritage. However, a growing ambivalence about the status and role of knowledge has formed the basis of a consciousness framed by the imperatives of risk management. This chapter discusses the way that the instrumental orientation of education reconceptualises the relationship between teacher and pupil, conceiving of teaching as a technical function rather than as a generational interaction between past, present, and future.

Keywords Relativism • Social construction • Teaching • University • Culture Wars

We have seen that debates about the purpose and nature of education in Britain during the interwar years revealed an ambivalence about which elements of the 'accumulated cultural heritage' should be passed on, and how this should be done. Through discussions about the teaching of English, the cultural elite of the time raised a number of questions about the role of national identity, and the kind of citizen that Britain's education system should create. Set against the backdrop of concerns about the nation's economic competitiveness and the health of its future workers and soldiers, this was a period in which the problem of generations met the problem of institutions: schools were charged with the responsibility for producing, *en masse*, the citizens of the future.

© The Editor(s) (if applicable) and The Author(s) 2016
J. Bristow, *The Sociology of Generations*,
DOI 10.1057/978-1-137-60136-0_3

Following the Second World War, the crisis in education took a some-what different turn—reflecting the extent to which knowledge itself became more explicitly contested. This was most explicitly revealed by the 'Culture Wars', with their endless battles over truth claims, curriculum content, and institutional power structures. One outcome of the Culture Wars was that a dogmatic attachment to tradition was replaced by an equally dogmatic attachment to what Allan Bloom, in *The Closing of the American Mind* (1987), described as 'education of openness'—in his view, an act of grotesque generational irresponsibility:

> It pays no attention to natural rights or the historical origins of our regime, which are now thought to have been essentially flawed and regressive. It is progressive and forward-looking. It does not demand fundamental agreement or the abandonment of old or new beliefs in favour of the natural ones. It is open to all kinds of men, all kinds of life-styles, all ideologies. There is no enemy other than the man who is not open to everything. But when there are no shared goals or vision of the public good, is the social contract any longer possible? (Bloom 1987, p. 27)

The extreme relativism of the Culture Wars has been the subject of some powerful critiques, both for its dissolution of the past and its anti-intellectual consequences in the present. Karen Carr, Assistant Professor of Religious Studies at Lawrence University, describes the consequence of post-modernism as 'the banalization of nihilism'. Nihilism, she argues, has lost its connotations with crisis and creativity and become a 'shoulder-shrugging' acceptance of the generalised futility of the search for truth. When knowledge is perceived as merely an endless series of perspectives, it 'devolves into its antithesis: a dogmatic absolutism' (Carr 1992, p. 10).

From a generational perspective, the promotion of the kind of 'education of openness' described by Bloom, and the trajectory of post-modernism critiqued by Carr, raise a number of problems. In this chapter, we begin by noting that the sociology of knowledge has long enjoyed an uneasy relationship with the problem of relativism. The very attempt to understand how knowledge is constructed and transmitted assumes a questioning of received wisdom, truth, or fact. In the post-war period, this questioning of the meaning of knowledge and its role in the reproduction of social control or conflict came to form a significant part of the *Zeitgeist* of the 1960s, reflecting both the intellectual currents of that time and the wider social and institutional upheavals that informed them. This questioning became part of the generational subjectivity of the students of the 1960s, who would become the teachers of the 1970s and 1980s. 'The revolting students of the 1960s are the revolting teachers of today, reproducing themselves by teaching as received wisdom what they furiously asserted against the wisdom received from their own teachers,' wrote Colin Welch in the *Spectator* (cited in Edgar 1986).

As we discuss below, the ideas that informed the generational consciousness of the 'Sixties generation' were significant, in promoting particular views about the role of education—summed up in the sentiment that 'the pedagogical is political'. Yet in examining the crisis of teaching in the present day, the legacy of the 'permissive Sixties' is not the full story. The relativisation of knowledge is often criticised by conservative thinkers, for whom the problem is the sociology of knowledge *itself*. As we have seen, there have been periodic attempts to push back against this trend through the construction of education policies that self-consciously seek to assert the authority of the past. Yet here, too, an ambivalence about knowledge reveals itself.

The self-conscious promotion of the cultural heritage by proponents of the 'neo-conservative traditionalist' model of curriculum development rests on foundations that are already uncertain about the ability of the older generation to transmit that legacy, and the wisdom of allowing young people to translate their heritage into their future. This leads to a relentless orientation away from the knowledge of the past and the authority of the older generation, and towards the political imperatives of the present day.

Relativism and the Sociology of Knowledge

Following Mannheim, we can accept that knowledge is contested and socially constructed whilst also acknowledging the importance of the cultural heritage. Mannheim's emphasis on the historical specificity of knowledge, its relationship to ideology, and significance of generations in the transmission and dynamic reconstruction of knowledge was an attempt to counter what he saw as the 'vague, ill-considered, and sterile form of relativism' that pertained with regard to scientific knowledge (Mannheim 1936, p. 17, p. 264).

For Mannheim, like Bloom, the loss of a shared system of social meaning was one of the most significant problems of his time. 'A society in which diverse groups can no longer agree on the meaning of God, Life, and Man, will be equally unable to decide unanimously what is to be understood by sin, despair, salvation, or loneliness,' he wrote in *Ideology and Utopia* (Mannheim 1936, p. 17, p. 264). His approach sought to understand and counter the fragmentation of meaning through theorising the process by which different perspectives on knowledge developed. 'Relationism does not signify that there are no criteria and rightness of wrongness in a discussion,' he wrote. 'It does insist, however, that it lies in the nature of certain assertions that they cannot be formulated absolutely, but only in terms of the perspective of the given situation' (Mannheim 1936, p. 283).

Mannheim's approach was not without its critics. 'Conservative critics attacked Mannheim as a subversive intellectual bent on undermining the dignity of mind and of spiritual values,' writes Remmling (1973), while 'left-wing social theorists ridiculed his sociology of knowledge as a decadent bourgeois game which, much like existentialism, questioned everything and attacked nothing' (Remmling 1973, p. 25). 'In an attempt to move away from the notion of a sociohistorical determinism, Mannheim dropped the term "relativism" and substituted "relationism"', write Curtis and Petras (1970). 'In summary, one finds few critics who disagree with W. Ziegenfuss's comment: "Mannheim's whole distinction between relationism and relativism is no more than a "play on words"' (Curtis and Petras 1970, p. 12).

While it is indeed debatable whether Mannheim's approach ultimately solves the problem of relativism, it seems unreasonable to dismiss this distinction as merely a 'play on words'. Mannheim's approach was motivated by a search for truth, and developed in the context of a keen appreciation of the importance of history and by an appreciation of wider social forces: both of which are markedly absent from post-modern theories, with their emphasis on language and multiple versions of reality.

Later developments within the sociology of knowledge, such as Berger and Luckmann's (1966) classic *The Social Construction of Reality* and the constructionist approaches to the study of social problems that developed from this, have navigated a similarly fine line. It is recognised that knowledge is dynamic, contested, and continually re-made; yet this does not mean that there is no such thing as knowledge, objectivity, or reality. 'It is precisely the dual character of society in terms of objective facticity *and* subjective meaning that makes its "reality *sui generis*"', state Berger and Luckmann (1991 [1966], p. 30).

As Best (2008) explains, social construction is not an arbitrary process, but one that is 'constrained by the physical world within which people find themselves'. While an imaginary society might construct 'all sorts of ridiculous meanings', in general 'the meanings people construct need to make sense of the world they inhabit' (Best 2008, pp. 11–2). The task of the sociology of knowledge is to understand the relationship between the (objective) world as it is and the (subjective) processes of meaning-making undertaken by its inhabitants.

Understanding the ways in which knowledge is socially constructed does not necessarily lead to relativism. Indeed, if it is pursued through a keen understanding of history and the wider context, social construction-ism is able to offer a far deeper and more nuanced account of the truth than knowledge claims based purely on established facts and received wisdom. Unfortunately, however, the trajectory of social constructionism over the latter part of the twentieth century tended either to privilege subjective meaning over objective facticity, or to merge with post-modernism in questioning the very notion of 'a fact'. In consequence, the project of education has often been conceptualised either as essentially political or as essentially meaningless.

TEACHERS: MEDIATING THE GENERATION GAP

In the early twenty-first century, the generational responsibility of the teacher is complicated by two powerful, and related, trends. The first is the degree to which teachers themselves are conscious of their responsibility to pass on the cultural heritage—or inclined, by their own formative experiences, to distance themselves from it. The second factor that has, over time, eroded the generational responsibility of teaching has been the politicisation of education and the erosion of the teacher's professional autonomy and status.

Understanding these trends requires grappling with the question of how teachers' own generational consciousness shapes the understanding of the world that they pass on to their students. It is sometimes assumed that conflicts over knowledge take place in one direction only, with the young deliberating over, and sometimes rejecting, that which the older generation consciously decides to teach. However, the process is more subtle and interactive—as Mannheim writes, 'not only does the teacher educate his pupil, but the pupil educates his teacher too'. The fact that '[g]enerations are in a state of constant interaction' mediates overt conflicts between the generations (Mannheim 1952, p. 301).

This continual mediation between the generations also means that, for older generations, 'holding the line' on what is known and what should be taught does not come naturally. What teachers know to be true is given, not only by the heritage that was passed down but also by the wider context that shaped their own fresh contact with that heritage, and by their experience of the present day, which in turn is mediated by their interaction with younger generations. 'What is consciously learned or inculcated belongs to those things which in the course of time have somehow, somewhere, become problematic and therefore invited conscious reflection,' explains Mannheim. 'This is why that inventory of experience which is absorbed by infiltration from the environment in early youth often becomes the historically oldest stratum of consciousness, which tends to stabilize itself as the natural view of the world' (Mannheim 1952, p. 299).

Where there is a mismatch between the 'natural' worldview of the teacher and that of his or her pupil, born out of the wider dynamics of the historical period that they inhabit, the process of conscious learning necessarily brings re-evaluation, and sometimes direct contestation. Mannheim explains this point by drawing further on his discussion of 'fresh contacts'. Youth are closer to the problems of the present day, and as such 'they are dramatically aware of a process of de-stabilization and take sides in it'. But at the same time, 'the older generation cling to the re-orientation that had been the drama of their youth'. Thus:

> From this angle, we can see that an adequate education or instruction of the young (in the sense of the complete transmission of all experiential stimuli which underlie pragmatic knowledge) would encounter a formidable difficulty in the fact that the experiential problems of the young are defined by a different set of adversaries from those of their teachers. Thus (apart from the exact sciences), the teacher-pupil relationship is not as between one representative of 'consciousness in general' and another, but as between one possible subjective centre of vital orientation and another subsequent one. (Mannheim 1952, p. 301)

The clash between two subjectivities forged in different times is what accounts for the dynamism of knowledge. It cannot be conceptualised as the passive transmission of simply 'what is known'; rather, the case for 'what is known' has to be made and a younger generation convinced of its truth. The extent to which the argument is won, or contested, is not given by the mere fact of generational change, although this fundamental underlying fact always provides the potential for contestation. It is affected by the wider social forces and intellectual trends operating at the time.

To put this another way: teachers can be told what to teach, but they cannot simply be instructed in what to *know*. Their own knowledge of the world is underscored by that which they learned as students—both as a result of 'conscious teaching' and the extent to which the ideas of their time 'seeped in'. When this knowledge clashes with the *Zeitgeist* of the present day, they will experience a 'generation gap' with their students. But when their knowledge of the world simply clashes with what they are being told to teach, the gap that opens up is not between the generations but between teachers and the role they are being instructed to perform.

This was the situation that framed the 'crisis of education' in the 1970s and 1980s, which came to a head over rows over curriculum control and accountability. Teachers, as one 'subjective centre of vital orientation', developed their knowledge of the world in the context of the intellectual currents and wider social events that have dominated the post-war world. As we explore below, the generational consciousness of the 1960s was one in which the 'accumulated cultural heritage' was openly contested and the role of the teacher represented as one of a political agent.

This orientation towards the 'pedagogical is political' situated teachers as change-makers, whose role as (adult) representatives of the past-present and proximity to (child) representatives of the present-future would be to encourage children to question the norms of society, rather than to reproduce them. As we discuss below, one effect of this was to position the teaching profession as a threat to politicians and to lay the basis for the further attempts to transform the role of the teacher from that of 'an individual professional trusted to make judgements about the needs of the community and individuals' to 'a technician employed to carry out national government policy' (Ward and Eden 2009, p. 102). Another effect was to contribute to a growing confusion about the distinction between adults and children, and the responsibility of the older generation in educating their young.

GENERATIONAL CONSCIOUSNESS AND THE PACE OF CHANGE

Mannheim emphasised the pace of change as a crucial factor in the extent to which generations become forged—first as an actuality and then as generation units, which would play a role in shaping and representing the *Zeitgeist*.

'Intellectual and cultural history is surely shaped, among other things, by social relations in which men get originally confronted with each other, by groups within which they find mutual stimulus, where concrete struggle produces entelechies and thereby also influences and to a large extent shapes art, religion, and so on,' he argued (Mannheim 1952, p. 285).

Mannheim perceived the 'tempo of social change' as a variable phenomenon, which, when accelerated, could lead to the 'formation of a new generation style, or... a new *generation entelechy*'. He explained:

> When as a result of an acceleration in the tempo of social and cultural transformation basic attitudes must change so quickly that the latent, continuous adaptation and modification of traditional patterns of experience, thought, and expression is no longer possible, then the various new phases of experience are consolidated somewhere, forming a clearly distinguishable new impulse, and a new centre of configuration. (Mannheim 1952, p. 309)

Over the twentieth century, moments of conflict between different generational subjectivities were expressed in a reaction against the institutions of education and in the ongoing contestation of knowledge. The character of this reaction is informed by wider social forces and the degree to which these promote an alternative vision of the ways in which the 'accumulated cultural heritage' should be shaped and transmitted.

The contestation over knowledge following the First World War can be most straightforwardly understood as deriving from a clash of ideologies, underpinned by class conflict and a generational consciousness framed by competing visions of the future. Clearly, the English curriculum promoted by the Newbolt Committee was informed in part by political considerations—most notably, the promotion of the spirit of Empire at a time when the authority of Empire was waning. The generational project that Lord Haldane considered at this time—preparing 'the future generation – morally, physically, and intellectually – to endure the strain' (*Manchester Guardian* 1916)—relied on the kind of ideas forged by the nineteenth-century 'age of ideology', which sat uneasily with the experience and outlook of the young people coming of age at that time. The bitterness expressed by the Generation of 1914, and the cynical detachment of young intellectuals of the 1920s (Marwick 1970), revealed a disjuncture between the cultural heritage and the uncertain, rapidly changing world in which educators and students alike now found themselves.

In period following the Second World War, the 'knowledge wars' took a somewhat different form. By 1960, when Daniel Bell proclaimed the 'end of ideology', politics derived 'from wholly different impulses to those of twenty years ago' (Bell 1960, p. 14). This, in turn, related to a number of significant social changes in America: changes that, in a slightly different form, would also become prominent in Britain. Bell detailed these as 'extraordinary' changes in the class structure, 'particularly in the growth of the white-collar class and the spread of suburbia'; the 'forced' expansion of an economy that was previously predicted to stagnate; expanded militarisation and the tensions of the Cold War; and 'a preoccupation with "self" and "status" that has brought to the fore not only psychoanalysis but the mirror of popular sociology' (Bell 1960, p. 13).

Following the Second World War, what Keniston (1971) described as 'The Speed-Up of Social Change' emerged as a dominant theme in the literature. In the 1960s, the pace and intensity of change was epitomised by a focus on the atomic bomb, which revealed its devastating power in Hiroshima and Nagasaki in 1945, becoming the basis of an immediate, and ongoing, existential threat during the Cuban Missile Crisis of 1962. The Baby Boomer generation, born and growing up during this period, was seen to be forged in part as a consequence of experiencing this world-defining (and potentially world-ending) technological change, in a wider context of social, institutional, and cultural upheaval.

One overt manifestation of this upheaval came in the form of the student protest movement that began on campuses in North America and spread across universities in Europe (Marwick 1999; Thomas 2002). There is a wealth of literature published in the early 1970s discussing the character and meaning of these protests, and their significance for the sociology of generations (see, for example, Keniston and Lerner 1971). This in part reflects the way that university campuses provided a physical and social environment conducive to the kind of politics and protest emblematic of the 1960s, which focused on culture and youthful rebellion. 'Since higher education is an increasingly central part of modern society, with larger proportions of the youthful population going on to college, the universities must be regarded as an increasingly important, perhaps the most important, agency of social change,' argued Goertzel (1972, p. 327).

Another, arguably more significant, feature of the 1960s was the way that the production, construction, and control of knowledge was explicitly contested, through the self-conscious creation of a 'counterculture' which set itself against the norms and values of liberal Western democracy. S.N. Eisenstadt's (1971) article 'Conflict and Intellectual Antinomianism' provides a compelling account of why the student protest movement emerged as such a critical and destabilising force in this context of higher education during this time and had such a powerful impact on generational consciousness. We review this in some detail below.

INTELLECTUAL ANTINOMIANISM AND THE ADVERSARY CULTURE

Eisenstadt describes 'intellectual antinomianism' as constituting 'an extreme manifestation of the tensions and ambivalence between intellectuals and authority which exist to a large extent in all human societies'. Modernisation, he argues, increases the tendency towards antinomianism, because of 'changes in the relations between the centers and the periphery' and because of 'growing structural differentiation in general, and of the spheres of intellectual, scientific and professional endeavour in particular'. However, 'perhaps the most important change related to these developments from the present point of view takes place in the social organization of the educational sphere' (Eisenstadt 1971, p. 72). With the onset of modernity:

> Education started to deal with the problems of forging new national communities and their common symbols, access to which tended to become more widely spread among different strata. At the same time, education began to serve increasingly as a channel of more general occupational, and allegedly achievement-based, selection. Moreover, the system of education tended to become more centralized and unified, thus assuring its permeation into wider strata of the society. (Eisenstadt 1971, p. 73)

This explanation brings together an acknowledgement of the role played by the institutions of education in reproducing social norms and structures and, as such, why education might become a focus for radical critique. But it is the symbolic role of the university that emerges as particularly significant in the context of the 1960s. This period, argues Eisenstadt, is characterised by 'a number of social and cultural contradictions and discontinuities' which have spread through society and focused increasingly on 'society's central symbols', most of which are to do with 'the tension between the premises of plenitude, full participation inherent in the symbolism of modernity, and the various structural limitations on the realization of these premises' (Eisenstadt 1971, p. 73).

In the cultural field, the most important development 'has been the transfer of emphasis from the creation of and participation in future-oriented collective values to the growing institutionalization of such values'. Eisenstadt explains further:

> This has been closely related to a very important shift in the whole pattern of protest in modern societies. Here, as in so many other cases, when much of the initial charismatic orientation and many of the goals have indeed become—through attainment of political independence, broadening of the scope of political participation, revolutionary changes of regime, development of welfare state policies, and the like—at least partially institutionalized, they give rise to new processes of change, to new series of problems and tensions, and to new foci of protest. (Eisenstadt 1971, p. 74)

The institutionalisation of dissent is associated with 'a marked decline of ideology in the traditional nineteenth- and early-twentieth-century sense, and a general flattening of traditional politico-ideological interest. This decline, in turn, has been connected with the growth of the feeling of spiritual or cultural shallowness in the new social and economic benefits accruing from the welfare state or from the "consumer society" ' (Eisenstadt 1971, p. 74). One consequence has been the 'breakdown of continuity in the historical consciousness or awareness', which Eisenstadt explains as follows:

> It is not only that the new generations have not experienced such events as the Depression or the two World Wars, which were crucial in the formation of their parents. What is more significant is that, probably partly due to the very process of institutionalization of the collective goals of their parents on the one hand, and their growing affluence on the other, the parent generation failed to transmit to the new generation the significance of the meaning of these historical events. The very emphasis on the new goals has increased a tendency to stress the novelty of the world created by the parents—a tendency taken up and reinforced by the younger generations. (Eisenstadt 1971, p. 75)

What Eisenstadt is describing here is the experience of rapid social change in a context where previous forms of meaning-making, framed by ideology, are seen to be exhausted. The effect of this is not limited to the political sphere: it changes the framing of recent history, as the older generation becomes unsure how to transmit its understanding of these events to their children. The result, argues Eisenstadt, is 'a whole series of structural and symbolical discontinuities', which have 'very often tended to culminate in a crisis of weakening of authority—evident in the lack of development of adequate role-models, on the one hand, and the erosion of many of the bases of legitimation of existing authority, on the other' (Eisenstadt 1971, pp. 75–6).

The destabilisation of adult authority, deriving from the struggle to make and transmit meaning about past events, meant that the generational consciousness developed by the young people who came of age during this time was one in which 'the possibility of linking personal transition' to 'societal and cosmic time' became further weakened. Eisenstadt explains:

> In general, these developments have depressed the image of the societal and cultural future and have deprived it of its allure. Either the ideological separation between present and future has become smaller or the two have tended to become entirely dissociated. Out of the first of these conditions has grown what Riesman has called the cult of immediacy; out of the second, a total negation of the present in the name of an entirely different future—both, in principle, totally unrelated to any consciousness of the past. (Eisenstadt 1971, p. 76)

This uncertainty about the future, born out of an inability to give meaning to the past, meant that the radical protest movements of the 1960s oriented themselves towards goals that were in many ways the opposite of those sought by 'the older, classical movements of protest of early modernity'. Whereas 'the major social and national movements... tended to assume that the framework and centers of the nation-state constituted the major cultural and social reference points of personal identity and that the major task before modern societies was to facilitate the access of broader strata of the society to these centers', the countercultural movements 'are characterized by their skepticism toward the new modern centers, by their lack of commitment to them, and by their tendency toward a lack of responsibility to the institutional and organizational frameworks of these centers' (Eisenstadt 1971, pp. 76–7).

The antinomianism that Eisenstadt identified as characteristic of this period had a particular importance for the university. Here, 'the social and cultural orders tend to become more salient and articulated' than in other institutions: the university has 'tended to become the major focus of the legitimation of a modern social order, and the attack on it indicates not only dissatisfaction with its own internal arrangements or even with the fact that it serves also as one mechanism of occupational and meritocratic selection' (Eisenstadt 1971, p. 68).

The attack on the university fundamentally 'emphasizes the denial that the existing order can realize these basic premises of modernity: to establish and maintain an order which could do justice to the claims to creativity and participation in the broader social order, and to overcome the various contradictions which have developed within it from the point of view of these claims'. This denial is 'often shared and emphasized by many of the faculty itself', which evinces some of the 'guilt feelings... of the parent generation in general and of the intellectuals among them in particular' (Eisenstadt 1971, p. 78). Thus:

> It is perhaps in the attack on the university that the new dimension of protest—the negation of the premises of modernity, the emphasis on the meaninglessness of the existing centers, and the symbols of collective identity—becomes articulated in the most extreme, although certainly not necessarily representative, way.
>
> It is also here that the basic themes of youth rebellion become very strongly connected with those of intellectual antinomianism. It is here that the rebellion against authority, hierarchy, and organizational framework, directed by the dreams of plenitude and of permissive, unstructured creativity, tends to become prominent—especially as the university serves also as the institutional meeting point between the educational and the central cultural spheres of the society. (Eisenstadt 1971, p. 78)

What Eisenstadt was describing in 1971 was the way that the university had become a site for the cultivation of dissatisfaction and dissent. This was a consequence of the growth of what Daniel Bell (1972) has described as an 'adversary culture', where cultural institutions, which previously operated to support the prevailing social and cultural norms and values, came to orient themselves against them. In the present day, the image of the 1960s is often framed in terms of the counterculture—a flabbier, and more generalised, rejection of capitalist institutions and values, associated with young people and the 'drop-out' hippie scene.

But following Eisenstadt, we can see that the roots of this cultural rejection go much deeper than the idealistic impressionism of 1960s' youth. It was the older generation's inability to give meaning to its past that framed the meaning that young people derived from their experience of the present. And it was the generational consciousness formed by students of the 1960s that framed the ways in which the cultural heritage was transmitted to children of the 1970s and 1980s.

Teachers as Agents of Change

The generational subjectivity forged by the 'Baby Boomer' generation, which came of age during this time of cultural and political upheaval, was forged less by its relationship with grand narratives and social alternatives, than by a sensibility that the project of history-making lay within personal action and experience. This has been described by Whalen and Flacks as the New Left vision of 'the permanent fusion of the everyday and history'—the sentiment that 'the personal is political and vice versa' (Whalen and Flacks 1989, pp. 9–10).

The complex and important elements that made up the turn towards 'the personal is political' is discussed in greater depth elsewhere (see Bristow 2015; Furedi 2014). Here, the aim is simply to note the radical implications that this had for the project of education. As indicated above, post-ideological forms of protest tended to focus on culture and the symbolic institutions of modernity, and focused on fostering a culture of scepticism and dissent. History-making became individualised and presentised: the scope of 'changing the world' was seen to be located within what individuals thought and did in the here and now.

In this context, the education of children came to be seen as one way in which change could be effected. By encouraging children to question the norms and values enshrined in society's accumulated cultural heritage, proponents of the idea that 'the pedagogical is political' sought to foster in the younger generation an aspiration to challenge these norms and values in the present day. Children's formative experiences would not be based on ideas about what was right with the world but on an empathetic engagement with what was wrong with it.

From a generational perspective, the politicisation of education in this way is highly problematic. When education is conceived as a political endeavour, children are engaged with less as new participants in the cultural process who need to be provided with a foundation for their world than as already-adults, whose mission is to engage with and solve contemporary problems. The unintended result of positioning children, prematurely, as change-makers is to anticipate and thwart their fresh contact with that which has gone before. The focus of what is learned never moves away from the urgency of the immediate: rather than bridging the past and the present, education works to keep children trapped in the present day.

As Furedi (2009) argues, '[e]ducation needs to be insulated from politics if it is to carry out the transaction between the generations responsibly'. He explains, following Arendt:

> Politics assumes a relationship of equality between participants; education is founded on the assumption that children need to be treated differently from adults. Most fundamentally, the difference between the mature and the immature, the old and the new, the teacher and learner, assumes a relation of inequality that education seeks to transcend. When children are not educated there can be no dialogue between equals, only an attempt at indoctrination. (Furedi 2009, p. 52)

By conceiving of education as a project of change-making, the notion that 'the pedagogical is political' privileged the subjectivity of the teacher over the generational responsibility required by education. In laying open the 'accumulated cultural heritage' to critique before children have gained access to this heritage as it is, the teacher privileges their own subjectivity over and above the authority of the past, and over and above the 'subjective centre of vital orientation' represented by their pupils. An approach that seems to be child-centred and future-oriented is, in fact, indoctrination into the 'cult of immediacy'.

It should be recognised that the radical educators who came out of the 1960s were not, by and large, attempting to indoctrinate children, nor to 'strike from their hands their chance of undertaking something new' (Arendt 2006, p. 193). Indeed, many had the opposite aim: to counter a narrow and prescriptive narrative of the past and encourage a commitment to open enquiry that was a part of their own formative *Zeitgeist*. This aim, in turn, was born out of a longer-running debate about the purpose of education. As the US educator and cultural critic Neil Postman explains:

> [O]ur citizens believe in two contradictory reasons for schooling. One is that schools must teach the young to accept the world as it is, with all of their culture's rules, requirements, constraints, and even prejudices. The other is that the young should be taught to be critical thinkers, so that they become men and women of independent mind, distanced from the conventional wisdom of their own time and with strength and skill enough to change what is wrong.
>
> Each of these beliefs is part of a unique narrative about what it means to be human, what it means to be a citizen, what it means to be intelligent. And each of these narratives can be found in the American tradition. (Postman 1996, p. 60)

But precisely because 'the spirit of the age' is forged in a particular time, it cannot be simply transplanted on to a different historical moment, inhabited by a generation forged by different experiences, problems, and battles. As the '1960s generation' continued to 'cling to the re-orientation that had been the drama of their youth' (Mannheim 1952), intellectual and institutional developments presented different problems. In this vein, Postman continues, when presented with the 'two contradictory reasons for schooling' outlined above:

> An author may think it necessary to subordinate one to the other—or vice versa—depending on what seems needed at a particular time. That is why, having co-authored *Teaching as a Subversive Activity*, he might later on write *Teaching as a Conserving Activity*. (Postman 1996, p. 60–1)

Teaching as a Subversive Activity was written by Postman in 1969 and *Teaching as a Conserving Activity* in 1979: reflecting the need for educators to remain sensitive to developments in education, what is lost when the pendulum swings, and where criticism may need to be re-directed.

A similar sensitivity was demonstrated by the British sociologist of education Michael Young, whose 1971 book *Knowledge and Control* drew attention to the power relations that informed school curricula and stands within the tradition of critical theory that became influential at this time. Yet in subsequent years, Young became increasingly uncomfortable with the way in which 'constructivist' accounts of the curriculum were tending in a post-modern direction, promoting a destructive relativism about academic subjects and denying children their 'entitlement to knowledge'. 'As a consequence, the"new" sociology of education that began... with a radical commitment to truthfulness, undermined its own project by its rejection of any idea of truth itself,' he wrote later, making the case for 'Bringing Knowledge Back In' (Young 2008, p. 199)

Young's response to the present-day crisis of education is to retain a commitment to subject knowledge, and the professionalism of the teacher. In *Knowledge and the Future School*, Young and Lambert (2014) argue that subjects remain 'the most reliable tools we have for enabling students to acquire knowledge and make sense of the world', and thus must be privileged over instrumental goals and generic skills. They make the case for a school curriculum that 'is not arbitrary or responsive to any kind of challenge; it is bounded by the epistemic rules of the particular specialist communities', yet which, unlike the conservative-traditional approach, 'does not treat knowledge as "given" but fallible and always open to change through the debates and research of the particular specialist community' (Young and Lambert 2014, p. 67)

From the perspective of the transaction between generations, the reorientation towards subject knowledge has great merit. To accept that knowledge is socially constructed means acknowledging that political and instrumental considerations can come to inform the kind of knowledge that is transmitted to the younger generation. This is why the content of the curriculum needs to be protected from intervention by politicians and policymakers, and regarded as the preserve of those who know their subject and can distinguish important new developments from passing fads. But this in turn requires that educators are considered to be professionals—subject specialists who are considered capable of, and responsible for, passing knowledge on to the younger generation. Trends of the past three decades have undermined this role.

Risk Consciousness and Teachers as Technicians

The 'de-professionalisation' of teaching has been widely noted and critiqued. It is often seen as a directly political attack by Thatcher's Conservative government of the 1980s against 'lefty' teachers, who were perceived as 'transmitters of the legacy of permissiveness (and the language of unearned rights) to the incapable and the immature' (Edgar 1986). Attacks on belligerent teachers' unions and the wider educational bureaucracy are seen to continue in Michael Gove's battle with 'the Blob'.

Yet as discussed in the previous chapter, politicians' attempts to challenge the autonomy and influence of the teaching profession go back to Labour Prime Minister Jim Callaghan's Ruskin College speech and formed an integral part of the reforms made by the New Labour government's reforms under Prime Minister Tony Blair. Ward and Eden summarise these developments as follows:

The period from the 1980s saw the government wrest control and definition of the curriculum from the professionals, and largely to succeed with the National Curriculum and testing. The late 1990s saw a narrowing of the primary curriculum with a re-emphasis on literacy and numeracy. Recent years have brought some loosening of the curriculum in state schools with more choice for secondary pupils, more 'creative' subjects and the encouragement of curricular activities in the academies. However, while the statu-

tory curriculum may have become less tightly controlled, what is taught
in schools is now defined by the state through government guidance and
direction, and closely monitored through its agents, the QCA and Ofsted.
(Ward and Eden 2009, p. 82)

While teachers' unions have complained about attacks on their professional status, much of the reaction against increasing centralised control
over what to teach and how to teach it has taken the form of complaints
about pressure, workload, and paperwork. Even during the bitter conflicts over the introduction of the National Curriculum in the 1980s,
concerns about professional autonomy were only one part of the complaint. 'Plagued by conflict and discord, the first years of the National
Curriculum saw various levels of discontent among professionals,' write
Ward and Eden (2009, p. 75). 'Some welcomed the prescription given to
them by the new curriculum, others resented it and there was continuing
complaint about the pressure both on teachers' workload and the effects
of testing on pupils.'

As the increased monitoring of teaching has become ingrained, it
is now argued that teaching is dominated by a 'culture of compliance'
(Ward and Eden 2009, p. 102), where teachers merely submit to latest
requirement to jump through particular hoops. To a certain extent, this
submission can be explained by the prosaic reality of needing to get on
with 'the job' when the battles have been lost; no doubt some teachers
still manage to incorporate a pragmatic adjustment to ticking the required
boxes while retaining their own sense of what needs to be taught and how
it should be done.

But underlying this culture of compliance, we can also see that the role
ascribed to the teacher today—neither as an authority on the past nor a
change-maker in the present, but as technical facilitator of current policy
imperatives—does not jar with the wider *Zeitgeist*. Indeed, it fits with ideas
that came to characterise the 'risk society' of the latter part of the twentieth century.

The sentiment that 'the pedagogical is political', associated with the
'Sixties generation' of teachers, presented the mission of the teacher as
effecting change in the here and now. This was problematic for the reasons
discussed above. It privileged the present over the past and saw the role
of education as inciting young people to change the world, rather than to
preserve it. Yet it was at least informed by an open orientation to change,

which saw the process of education, and the institutions of education, as important to knowing about the world and engaging in it.

In the second half of the twentieth century, however, there has been a shift in the way that rapid social change is conceptualised. Theories of 'risk society' assume that rapid change is a constant feature of life, to which individuals and social institutions alike must continually, reflexively adjust (Giddens 1991; Beck 1992). The resulting risk consciousness builds on the post-ideological orientation of the 1960s, in eschewing grand narratives and large-scale social solutions. However, unlike the vision of the future encapsulated in certain elements of the counterculture, which saw the future of history as a personal project, risk consciousness speaks to what Lasch (1984) has described as a culture of 'survival', in which the goal stops at the ability to manage one's personal life in an uncertain present.

The hope that political action will gradually humanize industrial society has given way to a determination to survive the general wreckage or, more modestly, to hold one's own life together in the face of mounting pressures,' writes Lasch in *The Minimal Self: Psychic Survival in Troubled Times*. 'The danger of personal disintegration encourages a sense of self-hood neither "imperial" nor "narcissistic" but simply beleaguered'. One consequence of this beleaguered selfhood is 'a kind of emotional retreat from the long-term commitments that presuppose a stable, secure, and orderly world.' (Lasch 1984, p. 16)

Regarding Lasch's thesis through the prism of the sociology of generations gives an important insight into some of the distinct tensions that frame the transmission of cultural heritage in the early twenty-first century. In a risk society, the sensibility of ceaseless change interacts with a belief that such change is detached from conscious human action, in the form of political causes or parties. At an individual and an institutional level, the focus is less on how people effect social change, than how people manage and mediate risk.

Risk consciousness has come to frame the outlook of the younger, 'Millennial' generation. In their discussion of 'the 9/11 generation', Edmunds and Turner (2005) argue that the terrorist attacks on the World Trade Center in 2001 'may form the immediate source of a new global generation, brought into being through developments that go back to the 1960s'. They explain these developments largely in terms of the growing globalisation of experiences and protests that has been made possible through mass travel and the media, which have enabled individuals from diverse geographical locations to experience a traumatic historical event as a common, generation-defining experience.

The defining feature of the 9/11 generation, argue Edmunds and Turner, is likely to be fear:

> The New York attacks could create a '9/11 Generation' that will be conscious of the negative effects of terrorism on their life-chances (for travel, urban security, global employment, civil liberties, national identity and relationship to religious movements and the Third World), thereby dividing them from the 1960s generation which experienced the global world, especially after the Cold War, as an open space. This new global generation is likely to be less complacent than the generation that preceded it. (Edmunds and Turner 2005, p. 571)

The effect of the 9/11 attacks, in this respect, was to concretise many of the fears and anxieties that lie behind risk consciousness. For the generation coming of age at the turn of the Millennium, the defining event of their youth appears as a problem of open space, open borders, and a surfeit of freedom. This feeds into the sentiment that change is something to be feared and risk a problem to be managed. The future is not only unknown, but unknowable.

It is important to stress that 9/11 did not create a generational consciousness that is grounded in fear. As we have discussed, risk consciousness has its roots in a wider crisis of knowledge that developed in Western societies in the second half of the twentieth century. The fact that 9/11 elicited the kind of generational response described by Edmunds and Turner is not a consequence of the attacks themselves: in a different time, framed by different ideas, these same horrific actions would not have led to an assumption of closed possibilities and febrile unease. But in concretising a set of pre-existing anxieties, the events of 9/11 and the response to them have given greater definition to the problem of generations in an era dominated by apprehension.

CONCLUSION

The trends described above have important consequences for the project of transmitting the cultural heritage. With its emphasis on constant change and the mediation of risk, today's society tends towards a dismissive approach to history. The past is indeed seen as 'another country', with wholly different rules, norms, and institutions; and the extent to which our understanding of the past can help to understand and shape the present is openly questioned.

When it comes to educating the young, discussions about the importance of tradition or historical knowledge have to contend with the idea that an appreciation of the past is irrelevant, even problematic, as it distracts the younger or future generations from grappling with the allegedly wholly different problems of today. The role of the teacher is re-cast in technical terms, as a facilitator whose aim is to assist young people in coping with the urgent demands of the present day.

Even when politicians assert the importance of transmitting the cultural heritage, they undermine it, through subjecting teachers to forms of bureaucratic management that seek to regulate their interaction with their students. Whether the issue is control over the content of the curriculum or the monitoring of teaching practice, the effect is the same. The clash of generational subjectivities that is inherent within the process of teaching, and what accounts for the dynamism of this relationship, is disrupted and flattened out.

The disruption of interaction between the generations, via mechanisms that seek to regulate what is passed on and how this is done, is not confined to teachers in the sphere of formal education. While successive education policies have played an important part in undermining teachers' professional role and status, they have been effective because they chime with a wider ambivalence about adult authority and the extent to which adults can be entrusted with the responsibility to transmit knowledge about the ways of the world to younger generations. This is the subject of the next chapter.

REFERENCES

Arendt, H. (2006 [1961]). *Between past and future: Eight exercises in political thought*. London: Penguin Books.

Beck, U. (1992). *Risk society: Towards a new modernity*. Thousand Oaks: Sage.

Bell, D. (1960). *The end of ideology*. New York: The Free Press of Glencoe (Macmillan).

Bell, D. (1972). The cultural contradictions of capitalism. *Journal of Aesthetic Education, 6*(1/2) Special Double Issue: Capitalism, Culture and Education, 11–18.

Berger, P., & Luckmann, T. (1991 [1966]). *The social construction of reality: A treatise in the sociology of knowledge*. London: Penguin.

Best, J. (2008). *Social problems*. New York/London: W.W. Norton & Company.

Bloom, A. (1987). *The closing of the American mind: How higher education has failed democracy and impoverished the souls of today's students*. New York: Simon and Schuster.

Bristow, J. (2015). *Baby boomers and generational conflict*. Basingstoke: Palgrave Macmillan.

Carr, K. (1992). *The banalization of nihilism: Twentieth-century responses to meaninglessness*. New York: State University of New York Press.

Curtis, J. E., & Petras, J. W. (Eds.). (1970). *The sociology of knowledge: A reader*. London: Gerlad Duckworth and Co. Ltd.

Edgar, D. (1986, July 7). It wasn't so naff in the 60s after all: The Conservative Party's assault on the legacy of the 1960s. *Guardian*.

Edmunds, J., & Turner, B. S. (2005). Global generations: Social change in the twentieth century. *The British Journal of Sociology, 56*(4), 559–577.

Eisenstadt, S. N. (1971). Generational conflict and intellectual antinomianism. *Annals of the American Academy of Political and Social Science, 395* Students Protest, 68–79.

Furedi, F. (2009). *Wasted: Why education isn't educating*. London/New York: Continuum.

Furedi, F. (2014). *First World War: Still no end in sight*. London: Bloomsbury.

Giddens, A. (1991). *Modernity and self-identity: Self and society in the late modern age*. Cambridge: Polity Press.

Goertzel, T. (1972). Generational conflict and social change. *Youth and Society, 3*(3), 327–352.

Keniston, K. (1971). *Youth and dissent*. New York: Harcourt Brace Jovanovich.

Keniston, K., & Lerner, M. (1971). Selected references on student protest. *Annals of the American Academy of Political and Social Science 395* (Students Protest), 184–194.

Lasch, C. (1984). *The minimal self: Psychic survival in troubled times*. New York/ London: WW Norton and Company.

Manchester Guardian. (1916, July 13). The training of the nation: Lord Haldane and the after-war struggle.

Mannheim, K. (1936). *Ideology and Utopia*. San Diego/New York/London: Harcourt Brace Jovanovich.

Mannheim, K. (1952). *Essays on the sociology of knowledge*. Paul Kecskemeti (Ed.). London: Routledge & Kegan Paul Ltd.

Marwick, A. (1970). Youth in Britain, 1920–1960: Detachment and commitment. Journal of Contemporary History 5(1) Generations in Conflict: 37–51.

Marwick, A. (1999). The sixties: Cultural revolution in Britain, France, Italy and the United States, c.1958–c.1974. Oxford: Oxford University Press.

Postman, N. (1996). *The end of education: Redefining the value of school*. New York: Vintage.

Remmling, G. W. (Ed.). (1973). *Towards the sociology of knowledge: Origin and development of a sociological thought style.* London: Routledge & Kegan Paul.

Thomas, N. (2002). Challenging myths of the 1960s: The case of student protest in Britain. Twentieth Century British History, 13(3), 277–297

Ward, S., & Eden, C. (2009). *Key issues in education policy.* Thousand Oaks: Sage.

Whalen, J., & Flacks, R. (1989). *Beyond the barricades: The sixties generation grows up.* Philadelphia: Temple University Press.

Young, M. (1971). *Knowledge and control: New directions in the sociology of education.* London: Macmillan.

Young, M. (2008). *Bringing knowledge back in: From social constructivism to social realism in the sociology of education.* Oxford/New York: Routledge.

Young, M., & Lambert, D. (2014). *Knowledge and the future school: Curriculum and social justice.* London: Bloomsbury.

CHAPTER 4

'Safeguarding', Child Protection and Implicit Knowledge

Abstract Much of children's knowledge of the world comes not from formal education but from implicit, everyday interactions between the generations, within the family and the community. This chapter discusses how the need to protect and socialise children is gradually devolving from a generalised generational responsibility into a bureaucratic function that seeks to distance children from the adult world, encapsulated in the language of 'safeguarding'. In this regard, the dynamic interaction between generations is rationalised, and flattened out.

Keywords Child abuse • Social work • Family • Every Child Matters • Safeguarding • Eating

This essay has situated the problem of generations within the wider problem of knowledge. Previous chapters have discussed how ambivalence about the accumulated cultural heritage has encouraged a process whereby teachers are placed at a distance from their students, and are increasingly cast as facilitators of skills, rather than mediators of knowledge. In this chapter, we discuss how the underlying crisis of knowledge in the present day destabilises the most fundamental feature of the generational transaction: the idea that adults have responsibility for the care of children.

Tensions surrounding the integration and socialisation of younger generations have existed throughout history, and emerged as particularly acute

© The Editor(s) (if applicable) and The Author(s) 2016   67
J. Bristow, *The Sociology of Generations*,
DOI 10.1057/978-1-137-60136-0_4

with the development of industrial society. While historical, sociological, and psychosocial accounts differ in their explanations for why these tensions emerged, most agree that the construction of childhood as a distinct phase of life is a feature of modern societies, which brings with it the question of how the transition from childhood to adulthood is managed (Ariès 1996; Cunningham 2006; Gillis 1974; Guldberg 2009; Postman 1994). Likewise, as kinship bonds become more fluid and less central to economic and social arrangements, questions emerge at the other end of the life course, to do with the engagement and care of elderly people (Jacoby 2011; Pilcher 1995; Phillipson 2013).

Acknowledging the historical 'problem of generations' is important in understanding the extent to which twenty-first-century Anglo-American society experiences, around the question of intergenerational contact, both a more developed version of an old problem and dynamics that appear novel to this era. The intimate relationship between knowledge and the people charged with transmitting that knowledge—teachers, parents, and other representatives of 'the older generation'—means that the contemporary problem of generations often tends to appear, first, as a crisis of interpersonal relations and institutional arrangements. In recent years, this has been exemplified by the mechanisms that have been adopted to prevent the abuse of children by their elders.

This chapter explores the development of the perception that children are 'at risk' from the adults that surround them—whether these be parents, 'other' adults in the community, or official child-carers, such as teachers or social workers (Best 1990; Hunt 2003; Jenkins 1998). We discuss the way that the child abuse paradigm has gradually expanded to encompass, not just incidences of suspected demonstrable harm to children, but much wider ideas about sub-optimal parenting (Munro 2007; Parton 2006). This has resulted in an increasing bureaucratisation of intergenerational contact, exemplified by the systems of surveillance and regulation known as 'safeguarding'. Such systems seek the explicit management of interactions between the generations, and frame the knowledge and practices of older generations as outdated, unhealthy, and dangerous.

CHILD ABUSE AS METAPHOR

A scandal over 'historic child abuse' has recently engulfed significant sections of the British political and cultural elite. The trigger for this was revelations about the entertainer Sir Jimmy Savile, who died in 2011 at the age of 84. Savile was posthumously accused of rape, indecent assault, and inappropriate behaviour with numerous children and older teenagers from his early career in the 1960s onwards. As more stories came to light about Savile's behaviour and the authorities' failure to investigate complaints at the time, an official inquiry was begun into suspected cover-ups of cases of 'historic' child abuse.

From a sociological point of view, the Savile scandal reveals a number of important cultural trends (Furedi 2013). The allegations made against Savile were made after his death, and based on events that allegedly took place some decades ago, primarily in the 1970s, by complainants now in middle age or older. As investigations into the scandal progressed, encompassing more high-profile individuals and a wider historical period, what emerged was the contestation of memory: not only individuals' memories of particular events but historical memory. Official institutions rapidly organised a response to a perceived *generational* threat, considered to stretch far beyond the reprehensible actions of particular individuals in a particular time and place, and beyond institutions directly concerned with the care of children to government circles, and cultural institutions such as the British Broadcasting Corporation (BBC). Launching the parliamentary inquiry into historical child abuse in 2015, Home Secretary Theresa May warned:

> We already know the trail will lead into our schools and hospitals, our churches, our youth clubs and many other institutions that should have been places of safety but instead became the setting for the most appalling abuse. However, what the country doesn't yet appreciate is the true scale of that abuse. (May 2015)

The sexual abuse of children, said May, is 'woven, covertly, into the fabric' of British society: and this official inquiry provided 'a once-in-a-generation opportunity to expose abuse and protect children in future' (BBC News Online 2015; May 2015).

Through this emphasis on the ubiquitous character of child sexual abuse, this phenomenon has become repositioned, not as an aberrant, abhorrent, deviant act, but as a normal feature of the 1960s and 1970s, which requires historical cleansing through condemnation in the present day. This highly symbolic approach indicates that the ongoing attempt to identify and prosecute abuses of the past is motivated by something far wider than a desire to bring justice to those who may have been victims of abuse thirty years ago. What is on trial here is our knowledge and experience of the past.

The phenomenon of child abuse is clearly real, in that some adults do physically or sexually assault some children. However, a large body of historical and sociological literature has illuminated the degree to which the form taken by the response to this phenomenon is *socially constructed*, in that it is shaped by ideas and agendas far wider than specific incidences of child harm. Sociologists situate the discovery of child abuse and the construction of the 'vulnerable' and 'at-risk' child within the framework of the construction of social problems, which attempts to grasp the way in which a particular social problem (in this case, child abuse) has been 'constructed' by the interplay of real events (the killing or harming of a child); the agendas of influential individuals or interest groups; the way stories have been reported by the media; and the way policymakers and official agencies have interacted with and responded to events, so that a problem becomes institutionalised (Best 1990). These events are, in turn, affected by wider cultural, political, and intellectual shifts, to do with conceptualisations of the family, and confusions surrounding adult identity and authority.

British journalists Barford and Westcott (2012) explain that the present-day 'hypervigilence' about child abuse is a relatively recent phenomenon. 'The 1980s, with some notable child abuse scandals, was in many ways the decade that child sexual abuse was "discovered"', they write:

> The Children Act 1989 became a massive landmark in child protection. People started to understand that abuse typically happened within families, but the threat to children in institutions from abusers also became clearer. (Barford and Westcott 2012)

This chronology of the discovery and the development of the child abuse problem is supported by sociological research into this question, which finds a very similar pattern in Britain and the USA (Best 1990; Jenkins 1998; La Fontaine 1998; Munro 2007; Parton 2006). Jenkins describes the years from 1976 to 1986 as 'the child abuse revolution', where a 'backlash' by campaigners for traditional moral values against the 'slide towards decadence' that they considered to be symbolised by the 1960s era and its 'tolerance of divorce, abortion, homosexuality, drugs, and sexual promiscuity' coincided with the agendas of radical feminists, for whom rape and child abuse provided a significant indictment of patriarchy and the male-dominated nuclear family (Jenkins 1998, p. 121, 125).

From the mid-1980s through to the present day, high-profile anxieties about child abuse have expanded on both sides of the Atlantic, coming to focus on child pornographers; 'predatory paedophiles' sometimes operating in 'paedophile rings'; social workers, teachers, residential care home workers, and other professionals working with children in a 'position of trust'; and adults in the community at large, who may gain unregulated access to children (Furedi and Bristow 2010; Piper and Stronach 2008; Webster 2005). Furthermore, child abuse has come to be seen as a problem that affects not only an expanding number of institutions but also an expanding range of behaviours. The concept of 'domain expansion' helps us to understand how a focus on a specific problem—for example, the 'battering' of babies by their mothers, which came to the fore in the early 1960s—gradually comes to include a wider range of problems, with, accordingly, a larger number of victims and perpetrators (Best 1990). The category of the *types of people* seen to pose as a threat to children has steadily widened, as has the *types of behaviours or interactions* that are framed as abuse.

The effect of the expansion of the child abuse frame has led to a reorganisation of generational relations that is novel to the twenty-first century. Adults have historically been positioned as the protectors of children; indeed, a key feature of adult identity has been the assumption that older generations should exercise a duty of care towards their young and that they can be trusted to do so. However, the growing perception that all adults are potentially a risk factor for children has led to a qualification of ideas about *de facto* generational responsibility by calls for its increasing restraint, and its replacement with formal, official, and bureaucratic forms of responsibility. 'Child protection' is posed as something that can only be offered by technical systems and disinterested experts; and the problems from which children are considered to need protecting are assumed to be features of adult life that were hitherto considered as 'normal'.

Every Child Matters

One example of domain expansion is provided in British policy by the development of the 'safeguarding agenda' that now dominates British policy on child welfare. Safeguarding is generally understood—and presented in official accounts—as the outcome of inquiries into specific cases of child murder. For example, the Every Child Matters Green Paper (HM Treasury 2003), which fundamentally reorganised the priorities of children's services in Britain, is presented as the outcome of Lord Laming's inquiry into the brutal neglect, abuse, and murder of eight-year-old Victoria Climbié in 2000 at the hands of her guardians (Laming 2003).

Yet there is a significant divergence between the findings of the Laming report into this horrific case and the proposals enshrined in Every Child Matters. As Parton writes, the Green Paper

> aimed to take forward many ideas about intervening at a much earlier stage in order to prevent a range of problems later in life, namely those related to educational attainment, unemployment and crime, particularly for children seen as 'in need' or 'at risk'. In this respect it aimed to build on much of the research and thinking [developed after 1997] and the policies introduced by New Labour in relation to childhood, where child development was seen as key and children were conceptualised primarily as future citizens. (Parton 2006, p. 139)

The five principles that inform the 'Every Child Matters' framework—be healthy; stay safe; enjoy and achieve; make a positive contribution; and achieve economic well-being—bear this point out. These generic ideas about children's well-being have strikingly little to do with what has generally been understood as child abuse, or child protection. While the five principles were popularised in relation to a high-profile case of child murder, they relate far more to pre-existing strategies relating to the socialisation of children, and the relationship between the family and systems of the state, and provide a further warrant for the increased monitoring of *all* families. 'The division in 2007 of the DfES [Department for Education and Skills] into the Department for Innovation, Universities and Skills (DIUS) and the Department for Children, Schools and Families (DCSF) was a sign that the government intended to move closer to the lives of children and their families,' observe Ward and Eden (2009, p. 7).

Every Child Matters proposed a wide-ranging system for the surveillance of all children, and the routine sharing of information between professionals and agencies—systems that, as child protection experts noted at the time, would have made no difference to the outcome of the Climbié case, and risked overloading social workers with information about children about whom there was no significant cause for concern (Munro 2003).

In a similar vein, the Vetting and Barring Scheme (VBS)—the system by which all adults in Britain who spend an amount of time working or volunteering with children have to undergo a police records check—is generally, and officially, presented as the outcome of the Bichard Inquiry that followed the murder, by school caretaker Ian Huntley, of ten-year-olds Jessica Chapman and Holly Wells in Soham in 2002 (Bichard 2004). But official proposals for a national system of 'safeguarding' were underway before this. The report *Safeguarding Children: A Join Chief Inspectors' Report on Arrangements to Safeguard Children*, published by the Department of Health in 2002 (before either the Laming or the Bichard reports), noted that '[t]he term safeguarding has not been defined in law or Government guidance', and provided a definition that, as Parton explains,

> demonstrated that not only was safeguarding the responsibility of a wide range of health, welfare and criminal justice agencies, who needed to work closely together and share information, but also that safeguarding was about *concerns about children and young people's welfare* as well as *risks of harm to children's welfare*. We thus have far more agencies identified as being responsible and a broad focus of what those responsibilities should be. (Parton 2006, p. 142. Emphasis in original)

As definitions of abuse have expanded beyond the cases of demonstrable, physical harm that were traditionally the concern of social workers to include a far broader raft of concerns about child welfare, and as the range of professionals considered necessary to monitor and intervene in family life has widened from social workers to include health professionals, teachers, and others connected to schools, the degree to which parenting has come under scrutiny has increased significantly. Ball (2013) includes 'parenting' as one of the 'current key issues' dominating the education debate and argues:

> The interventionary dimension of policy is invested with aspects of 'moralisation' and the importance given to civic 'responsibilities'. In these policy discourses, parents are key figures in regenerating social morality and lack of parental discipline is linked to problems of truancy, anti-social behaviour, offending and obesity. (Ball 2013, p. 200)

The policy focus on parents and parenting is explicitly linked to the ways in which policymakers perceive and manage the institutions of education. Ball explains:

> In Foucault's (1979) terms such policies enact 'dividing practices', procedures that objectify subjects (feckless parents) as socially and politically irresponsible. These are the 'others' of policy who need to be 'saved' from their uncivilised lives through expert 'interventions'. They are incapable of being responsible for themselves and their children. In effect this is the same model of governing as that applied to institutions, a certain sort of freedom is offered, a virtuous, disciplined and responsible autonomy that, if not taken up appropriately, provokes 'intervention', as is the case with 'failing' and

'underperforming' schools. We even see here further extension of 'contracting' as a way of representing relationships between institutions, between individuals and institutions, and between individuals one with another, in the form of 'home-school contracts'. (Ball 2013, p. 201)

Via the imperative of safeguarding, the expansion of the 'at-risk' category has been justified through an orthodoxy that sees an increased range and extent of risk factors purportedly affecting a child, and has brought together a number of agencies in monitoring everyday child-rearing practices. Thus, while a child two decades ago would be considered as 'at risk' only from neglect, sexual abuse, or physical harm, a child is now positioned as 'at risk' of failing to achieve his or her full potential or happiness, because of a presumed parenting deficit. The logical consequence of this orthodoxy is that all children are to some degree 'at risk', and surveillance by experts external to the family, the community, and the professions is needed to monitor the well-being of children as whole.

RISK AND REGULATION

Part of the explanation for the contemporary tendency to visualise child abuse as an increasing problem lies in the heightened level of risk consciousness that has come to shape parenting attitudes and practices in the modern era. When we attempt to understand the parental fears about child abuse at the hands of other adults, we should not assume that these fears arise from the *actual* prevalence of abuse or likelihood that their particular child will be abused; we should also recognise that the free-floating anxieties that parents have about their children become focused on particular hypothetical risks (Hunt 2003; Skenazy 2010).

Regulatory schemes that are launched to assuage such fears and increase children's safety tend to have the paradoxical effect of expanding and deepening fears. In the USA, this is exemplified by the establishment of 'community notification' statutes, popularly known as 'Megan's Law', which require that the authorities notify neighbours and schools 'of the presence of high-risk offenders within the community' (Jenkins 1998, p. 198). In Britain, the VBS, discussed above, introduced a system whereby any adult who wished to work or volunteer with children had to undergo a formal criminal records check before they were permitted to do so.

Both Megan's Law and the VBS could be seen as regulatory projects that focus on the risks posed to children by 'other' adults—a category that spans a wide range of adults, from those who have previously been convicted by the courts of abusing a child to those who simply have access to children, through community groups or their professional work. The effect of such schemes has been recognised to increase the individuation and isolation of parents; while the stated intention is to increase trust in those who have been officially 'licensed to hug', there are indications that they have the opposite effect (DfE, DH, HO 2011; Furedi and Bristow 2010; Lee et al. 2014). For example, as McAlinden (2010) notes in her study of the vetting of sexual offenders in Britain under the Safeguarding Vulnerable Groups Act 2006, '"hyper innovation" and state over-extension' in this area has resulted in 'exceptionally uncertain and unsafe policies', which lead, among other things, to 'unintended and ambiguous policy effects', and ultimately draw attention to 'the failure of the state to deliver on its self-imposed regulatory mandate to effectively manage risk' (McAlinden 2010, p. 25).

The hyper-regulation of adult-child contact thus results in neither safety nor reassurance. Parents are encouraged not to rely on spontaneous instincts of trust, and instead to demand official clearance; yet it quickly becomes clear that official regulation cannot prevent all cases of abuse from happening, leading to an increase in mistrust and a call for yet more regulation.

In a different era, it might be possible to see the increasing suspicion of 'other' adults as an indication of a cultural agenda to promote the idealisation of the nuclear family, and particularly to encourage women to focus on home and child-rearing. Over the post-war period, cultural attitudes towards women working outside the home have fluctuated according to the political and economic agendas of particular times, and this in turn has had a practical impact on the provision of childcare and the 'socialisation' of other forms of domestic work (Randall 2000; Riley 1983; Somerville 2000). More recently, sociological accounts of 'intensive parenting' have highlighted the extent to which the contemporary cultural obsession with rearing children according to the highest educational, behavioural and emotional standards has increased the degree of individuation experienced by mothers, in particular (Blum 2015; Hays 1996; Lee et al. 2014).

There is little doubt that a culture of intensive, individuated, and risk-averse parenting has reinforced a barrier between children and 'other' adults—that is, adults outside of the family, both those with professional responsibilities for children (such as teachers and social workers), and adults in the community at large. Over the past few decades, on both sides of the Atlantic, social workers have been frequently castigated for

their incompetence; care home workers have been accused of molesta-
tion, cruelty, and other forms of abuse; teachers are scrutinised for past
allegations of untoward behaviour; and nursery workers and childmind-
ers are increasingly monitored and regulated (Jones 2004; Murray 1996,
2001; Piper et al. 2006; Sikes and Piper 2010). The experience of the
past forty years has been a growing suspicion of contact between children
and non-family members, and an increasing tendency to formalise and
regulate these relationships, positing the child as increasingly 'at risk'
from 'stranger danger' or, in relation to adults in a professional role, an
'abuse of trust'.

However, the very trends that have caused anxiety about intergenerational contact outside of the family simultaneously affect relations within
the family. It is the very intensity of the parent child relationship that
makes this appear as a prime site for abuse—reflecting the transformation
of assumptions about intimate relationships, discussed in the next chapter.
Parents are situated both as the *only* people who care enough, and can be
trusted enough, to protect their own children from the myriad risks posed
by the outside world. At the same time, parents' unique relationship of
power over, and trust with, their own children, and the privacy that has
historically been accorded to the family, is seen to present an especially
powerful risk factor, in that abuse of a child by a parent is more likely to
have a more sustained and intimate character, and to go undetected or
unreported.

There is, in the present day, no presumption that *any* group of adults
will protect the children in their care—the presumption is that they may
pose a danger, and should be monitored accordingly. In this context, the
cause of 'child protection' is transformed from a generational responsibility, of adults towards children, into a disembodied, disinterested,
bureaucratic system. This is where the 'child protection expert' comes
to play a prominent role, over a body of children considered to be, *en
masse*, 'at risk'.

INTERGENERATIONAL ESTRANGEMENT AND THE PROBLEM
OF ADULT IDENTITY

In his 1977 book *Haven in a Heartless World: The family besieged*, the
American cultural historian Christopher Lasch (1977) analysed the gradual
invasion of the family by the combined forces of commercialisation, medi-
calisation and therapy culture. By the third quarter of the twentieth century,

he argued, '[t]he citizen's entire existence has... been subjected to social direction, increasingly unmediated by the family or other institutions to which the work of socialization was once confined' (Lasch 1977, p. 189).

For Lasch, the decline of authority within the adult community at large is both mirrored and exacerbated by the erosion of parental authority within the family. External agencies and cultural influences were taking on more and more aspects of the socialisation process, and the family, in turn, was subject to the rules and norms of these agencies and influences. In consequence, argued Lasch, '[r]elations within the family have come to resemble relations in the rest of society. Parents refrain from arbitrarily imposing their wishes on the child, thereby making it clear that authority deserves to be regarded as valid only insofar as it conforms to reason' (Lasch 1977, p. 174).

Lasch's analysis situates the growth of bureaucratic regulation of, and intervention in, the family within a broader context of the weakening of adult identity and authority. *Haven in a Heartless World* and Lasch's subsequent books *The Culture of Narcissism* (1979) and *The Minimal Self* (1984) discuss the therapeutic management of society as a process that is intimately connected to a wider sensibility of individuation and self-absorption. Adults are infantilised by the incursions on their authority within the home, and by the cultural trends that incite them to dwell on past experiences and relationships as the cause of their present problems, and to seek therapeutic guidance and affirmation in the art of living. This infantilisation makes it difficult for adults to exercise authority over children.

These insights have been developed by Furedi (2001), who suggests that the 'child-centred' ethos of twenty-first century society is better understood as 'child-obsessed'. 'An examination of our culture's preoccupation with the child suggests that this development is inseparable from some of the problems that afflict the world of adults,' he writes. 'It is the uncertainties which surround the meaning of adult identities that motivate many parents to put so much of their emotional capital into children' (Furedi 2001, p. 101). Highly sensible of their own vulnerability and infantilised by a culture of expert advice and support, adults' fragile sense of self-identity appears to be doubly threatened by the risks that seem to surround their children. The risks take many forms: health risks, environmental risks, and risks of damage to self-esteem and psychological wellbeing. Crucially, many of these risks are seen to emanate directly from the adult world—either in the form of the predatory individual abuser or in the more diffuse form of a harmful adult culture.

In this respect, the desire to protect one's children from risk has come to mean protecting them from adults, and adulthood. In recent years, the metaphor of 'toxicity'—the notion that modern life *itself* is poisoning future generations—appears to have gained momentum. Many of the anxieties about the effects upon children of the toxicity of the modern world are versions of the concerns that preoccupy adults—obesity, consumerism, celebrity culture, sexual confusion, fast food, and the rise in mental health problems, to name just a few examples. However, while many of these concerns are simply mapped onto children, the notion of toxicity also develops a subtly different form when it is presented self-consciously in relation to childhood. 'The speed of technological and cultural change has been so fast that we haven't really had time to think about it,' writes Sue Palmer, in her introduction to *Detoxing Childhood*. She continues:

> And since, for most adults, the changes have been generally welcome, we haven't really bothered to think. But it's now becoming clear that some aspects of modern life are seriously damaging our children. (Palmer 2007 p. 1)

The counterposition made here between the interests of adults (for whom changes have been 'generally welcome') and those of children indicates that what is 'toxic', in this view, is not a specifically dysfunctional form of behaviour or set of values, but the very process of attempting to socialise children. Adults are positioned as the vector for transmitting the toxicity of modern life to the next generation.

GENERATIONAL DISTANCING AND SOCIALISATION IN REVERSE

Notions of the child as 'sacred', 'innocent', or 'pure', and in need of protection from the apparent corruption of the adult world, have a long history, and do not themselves challenge the idea that children will grow into adults, or that this is desirable. What is distinctive about the contemporary period is that there exists a far greater ambivalence about both the capacity of adults to protect children from the excesses of the world outside the home and the desirability of socialisation into that world.

The linking of adulthood to negative values and modes of behaviour has become a feature of mainstream policy discourse, exemplified through the concepts of 'intergenerational transmission'—the 'intergenerational transmission of disadvantage' (d'Addio 2007; Office for National Statistics 2014), the 'intergenerational transmission of worklessness' (Schoon et al. 2012), or the 'intergenerational transmission of young motherhood' (Stanfors and Scott 2013), to name just a few examples. A common feature of all such claims is the promotion of the idea that the relationship between generations within the family is *de facto* problematic, as it becomes a vehicle for transmitting, as common sense, attitudes and behaviours that are considered either socially problematic or merely outdated.

The preoccupation with the generational transmission of negative attitudes has had two main policy consequences. The first, as discussed further in the next chapter, is the engineering of policy designed to position the expert or official more intimately within the family, as a competing pole of authority to the parent. This is exemplified by the recent wave of policy initiatives around early intervention, and the expanded role allocated to schools in relation to what previous eras considered to be primarily 'family matters' to do with emotional care and socialisation, such as advice about sex, bullying, and healthy eating. As Furedi explains, this approach has important implications for intergenerational relations, in that '[i]mplicitly, the direct socialisation of children into expert-derived values serves to distance the young from the old' (Furedi 2009, p. 105).

The second policy consequence is that which Furedi has termed 'socialisation in reverse', 'a phenomenon where children are entrusted with the mission of socialising their elders' (Furedi 2009, p. 89). Here, the socialisation of young people by their elders is not only bypassed—it is fed back into the generations from experts to adults via their children. In this respect, the meaning of socialisation is transformed (reversed) from the traditional view that adults need to rear their children to participate in the adult world to the idea that children need to teach their parents how to behave.

As Furedi's discussion of education notes, the idea that policymakers should look to the child as a way of instilling good values and behaviour in adults is not entirely new. For example, it is indicated by the influence enjoyed by the 'salvationist' view of children held by progressive educators in the early twentieth century, which held that children were innately good and that harnessing their spontaneous impulses, rather than attempting to shape and control them, could lead to the 'moral regeneration of the nation' (Furedi 2009, p. 95). But in the present day, children are routinely viewed, and used, as ciphers for expert guidance on appropriate attitudes and behaviours. It is not the moral virtue of the child that is seen as the element that can guide confused or corrupt adults to create a better world: rather, the positioning of the child in a credulous relationship with expert guidance is seen as an appropriate and effective locus of social control.

'Healthy Eating' and the Problem of Knowledge

One example of socialisation in reverse is provided by the preoccupation with an 'obesity epidemic' across the Western world and the use of campaigns aiming to change families' eating habits by targeting the children. As the weight of both the adult and child population has increased over recent decades, largely thanks to an abundance of cheaply available food, the health effects of 'obesity' have become a major public health concern. The precise link between weight and health problems, in the absence of other confounding factors, continues to be debated, as does the relationship between 'childhood obesity' and obesity in adults (Campos 2004; Lyons 2011). As the Australian writers Michael Gard and Jan Wright note, the 'obesity epidemic' is a socially constructed problem, which has a 'number of ingredients'—the use of scientific claims to find 'certainty in the face of uncertainty'; the entrenchment of assumptions about the causes and consequences of obesity by 'widely held popular beliefs'; and the reliance of the 'obesity epidemic' claim on 'a particular form of morality that sees the problem as a product of individual failing and weakness' (Gard and Wright 2005, p. 7).

Gard and Wright argue that 'an intellectually critical approach to the so-called "obesity epidemic"' is important because of the regulatory and cultural consequences that arise from a misdiagnosis of the problem. As they write:

> Faced with an 'epidemic', all manner of drastic measures are likely to be advocated and enacted by policy makers and others in a position of authority. If the label 'epidemic' *is* unwarranted, then it is possible that many of these measures will be unwise, unnecessary and wasteful of scarce resources, while others may turn out to be counter-productive or even harmful in unforeseen ways. (Gard and Wright 2005, p. 8; emphasis in original)

Such 'drastic' regulatory measures have been apparent in both Britain and the USA in recent years. In Britain, one noteworthy measure has come in the form of the routine monitoring of children's lunch boxes in schools, with many schools imposing outright bans on chocolate and fizzy drinks, accompanied by the public health campaign 'Change 4 Life', which aims to encourage children to take the healthy message home to their parents (NHS 2015). Politicians frequently use the alarmist rhetoric of the childhood obesity crisis to justify overtly authoritarian family policies. 'We don't want to tell people how to lead their lives, that is a Conservative principle,' stated Health Secretary Jeremy Hunt in 2015, quickly adding:

> We don't like a nanny state except when it comes to children. Children are allowed nannies and I think we're able to be a little bit more draconian when it comes to childhood obesity. (Spencer et al. 2015)

The explicit objective of healthy eating campaigns targeted at children is that of socialisation in reverse. It is hoped that by instructing children about the need to eat only 'good' foods, their parents will take note of what they should be feeding their children, and what they should be eating themselves. Yet the premise of this campaign is not one that sees children's impulses as the right ones—many of the foods targeted by such campaigns are those that children particularly like to eat, such as sweets and chips. Rather, children are positioned as the group most likely to have their behaviour influenced by expert guidance and the voice most likely to be listened to by parents. Children themselves are not the 'salvation' of adult diets, as left to their own devices they would primarily eat the 'bad foods'; they are merely the ciphers of regulation, in the form of expert advice.

The impact of the current campaigns against particular types of food upon physical health is the subject of ongoing contestation. What is evident, however, is that such campaigns have wider consequences for relations within the family and between generations. Food in modern society is not merely 'fuel' for the human body; it is implicitly bound up with a number of social and cultural aspects of daily life (Keenan and Stapleton 2010). The historical creation of the 'family meal' as a time for adults and children to be together was informed more by ideas about the conduct of family relations than it was by standards of nutrition (Gillis 1997), and the routine exercise of parental authority that goes alongside admonishments to 'eat your greens' or 'finish what's on your plate' represent a combination of concerns about balanced diets, the family budget, and obedience to parental authority. The giving of 'treats' to children, by grandparents can be seen to represent an older generation's desire to show that it knows what children like, and that—as elders—it has the freedom to provide the kind of indulgence that parents do not.

By inserting, via the child, the notion that food should be judged separately from its context, in terms of clearly polarised notions about what is healthy and what is not, recent healthy eating campaigns seek to disrupt tacit generational attitudes towards food and eating. This incites adults to question, or be questioned, about routine aspects of daily life— giving a packet of sweets to a grandchild or insisting that a child finish his or her dinner—and invites a constant deference to experts. This in turn fuels the sentiment that older generations are out of time and out of touch when it comes to rearing children and tacitly transmits the notion that they should hold back and leave the 'words of wisdom' to the experts.

CONCLUSION

The 'most important' aspect of the process of cultural transmission, Mannheim argued, is 'the automatic passing on to the new generations of the traditional ways of life, feelings, and attitudes' (Mannheim 1952, p. 299). In this respect, we can see that what appears as a problem of food, or a problem of parenting, is in fact a problem of knowledge. The obsession with children's bodies derives from a crisis of the adult mindset, in which food and eating practices, as symbols of the 'old ways', are automatically positioned as problematic, and only the expert guidance generated in the present day is seen to hold authority.

Healthy eating campaigns provide merely one example of the way in which the idea that older generations have a responsibility to educate and socialise the young about their world, in order to enable them to run this world when they are adults, is contradicted by a sentiment that sees children's detachment from this 'toxic' world as the very quality that imbues them with moral stature.

In this trend, we see a continuation of the decades-old anxiety about the 'generation gap', but an inversion of the way that the problem was posed historically. In previous eras, the resolution of generational tensions has been seen to lie within adult society; specifically, its ability (or inability) to use the promise of adult life to overcome the alienation of the disenfranchised, powerless youth and to encourage a better understanding of and engagement with adult society by the young.

For example, in the 1950s and early 1960s, sociologists and policymakers were preoccupied with the emergence of the 'teenager' as a distinct social group, which enjoyed a certain amount of leisure time and disposable income without being engaged in the discipline of work, and 'delinquency' emerged as a particular concern (Gillis 1974). But the resolution to this problem was seen to lie in the ability of adult society to transmit its values and authority in a clearer way. As John Barron Mays wrote, in his 1961 article about 'Teenage Culture in Contemporary Britain and Europe':

> The majority of those who rebel in this period would, given adequate support and firm but sympathetic leadership, adjust to their growing-up problems in socially acceptable ways. But the failure of older members of the community, especially of parents and educators, to give them adequate support, makes them temporarily easy victims for the illegal promptings of a handful of seriously maladjusted and emotionally disturbed instigators. (Mays 1961, p. 27)

Contemporary manifestations of the generation gap, by contrast, emphasise the 'outdated' character of the ideas and experience of older generations and seek to draw inspiration and guidance from the young. The norms and customs of family life, and the socialisation of children, are increasingly subject to the orthodoxy that the 'old ways' are dangerous and undesirable and that the only way to 'parent' is to follow the advice of the twenty-first century expert. One result of this is that older generations find themselves increasingly nervous about the ways in which they might spontaneously touch, discipline, feed, or educate children.

By situating these developments within the sociology of knowledge, we can see that what appears as a crisis of particular interpersonal relationships (an abusive father or out-of-touch grandparent) or specific institutional failings (incompetent social workers or 'inappropriate' teachers) has its roots in a more far-reaching uncertainty about the kind of world adults are able to create for their young, and their capacity either to integrate children into this world or protect them from it (Arendt 2006; Furedi 2009; Mannheim 1952). This reminds us that the problem of generations is properly situated, not within the private sphere of interpersonal relations but within the realm of cultural renewal. The focus on 'parenting' that now dominates the discourse about child-rearing, and the preoccupation with finding systems to ensure 'child protection' from harms resulting from undesirable forms of intergenerational contact, offers neither the cause nor the solution to the problem of generation in the twenty-first century.

As cultural and political institutions seek to diagnose uncertainty and disenchantment as problems of adulthood, and attempt to manage them through an increasing intervention into and manipulation of the lives of children, the relationship between generations becomes further confused. The crisis of adulthood achieves its most acute form in relation to adolescents and young adults, who are simultaneously framed as victimised by the adult world and as playing a role in the 'toxification' of childhood. It has its echo at the other end of the life course, where elderly people are framed as both 'to blame' for the problems of the adult world and irrelevant to making a continuing contribution to it, and as 'vulnerable' to abuse by adults of working age.

While the problem of generations is at source a cultural one, we have identified in this chapter some very real and practical consequences at the level of intergenerational relations. Both the formal and spontaneous processes of cultural transmission theorised by Mannheim are distorted by bureaucratic systems of risk management, which privilege the role of the disinterested expert over those individuals—parents, teachers, and adults within the community—who have a relationship and responsibility for the child. The job of 'child protection' is increasingly managed by databases and bureaucracies, where impersonality is seen as an important safeguard against corruptibility.

Impersonality, however, is fundamentally incapable of either rearing or protecting children. When child-rearing is seen, not as a generational responsibility but as the responsibility of the 'intensive parent' following expert guidance (Lee et al. 2014), this contributes to a heightened sense of individuation and detachment from members of the wider community. And when the parent is seen as a potential risk factor to his or her child, and the professional who engages with the parent also has his or her motives scrutinised, the overall effect is one of distance—between parents and children, parents and professionals, and parents and other adults.

REFERENCES

Arendt, H. (2006 [1961]). *Between past and future: Eight exercises in political thought*. London: Penguin Books.

Ariès, P. (1996 [1960]). *Centuries of childhood*. London: Pimlico.

Ball, S. J. (2013). *The education debate* (2nd ed.). Bristol: The Policy Press.

Barford, V., & Westcott, K. (2012, October 29). Jimmy Savile: The road to hyper-vigilance. *BBC News Magazine*. Accessed Available at: http://www.bbc.co.uk/news/magazine-20093812

BBC News Online. (2015, March 14). Child sex abuse is "woven into British society" – Theresa May. Available at: http://www.bbc.co.uk/news/uk-31885906. Accessed 9 Dec 2015.

Best, J. (1990). *Threatened children: Rhetoric and concern about child-victims.* Chicago/London: University of Chicago Press.

Bichard, M. (2004). *The Bichard Inquiry – Report.* London: The Stationery Office. Available at: http://dera.ioe.ac.uk/6394/1/report.pdf. Accessed 18 Dec 2015.

Blum, L. M. (2015). *Raising Generation Rx: Mothering kids with invisible disabilities in an age of inequality.* New York: New York University Press.

Campos, P. F. (2004). *The obesity myth: Why America's obsession with weight is hazardous to your health.* New York: Gotham Books.

Cunningham, H. (2006). *The invention of childhood.* London: BBC Books.

d'Addio, A. C. (2007). Intergenerational transmission of disadvantage: Mobility or immobility across generations? A review of the evidence for OECD countries. *OECD social, employment and migration working papers no. 52.* Available at: http://www.oecd.org/els/38335410.pdf. Accessed 18 Dec 2015.

Department for Education, Department of Health, Home Office (DfE, DH, HO). (2011, February). *Vetting & barring scheme remodelling review – Report and recommendations.* Available at: https://www.education.gov.uk/publications/eOrderingDownload/vbs-report.pdf. Accessed 18 Dec 2015.

Furedi, F. (2001). *Paranoid parenting: Why ignoring the experts may be best for your child.* London: Allen Lane.

Furedi, F. (2009). *Wasted: Why education isn't educating.* London/New York: Continuum.

Furedi, F. (2013). *Moral crusades in an age of mistrust: The Jimmy Savile scandal.* Basingstoke: Palgrave Macmillan.

Furedi, F., & Bristow, J. (2010). *Licensed to hug: How child protection policies are poisoning the relationship between the generations and damaging the voluntary sector* (2nd ed.). London: Civitas.

Gard, M., & Wright, J. (2005). The obesity epidemic: Science, morality and ideology. London/New York: Routledge.

Gillis, J. R. (1974). *Youth and history: Tradition and change in European age relations 1770-present.* New York/London: Academic Press.

Gillis, J. R. (1997). *A world of their own making: A history of myth and ritual in family life.* Oxford: Oxford University Press.

Guldberg, H. (2009). *Reclaiming childhood: Freedom and play in an age of fear.* London/New York: Routledge.

Hays, S. (1996). *The cultural contradictions of motherhood.* New Haven/London: Yale University Press.

HM Treasury. (2003). *Every child matters.* London: The Stationery Office. Available at: https://www.gov.uk/government/uploads/system/uploads/ attachment_data/file/272064/5860.pdf. Accessed 9 Dec 2015.

Hunt, A. (2003). Risk and moralization in everyday life. In R. Ericson & A. Doyle (Eds.), *Risk and morality* (pp. 165–192). Toronto: University of Toronto Press.

Jacoby, S. (2011). *Never say die: The myth and marketing of the new old age.* New York: Pantheon Books.

Jenkins, P. (1998). *Moral panic: Changing concepts of the child molester in modern America.* New Haven/London: Yale University Press.

Jones, A. (2004). Risk anxiety, policy, and the spectre of sexual abuse in early childhood education. *Discourse: Studies in the Cultural Politics of Education, 25*(3), 321–334.

Keenan, J., & Stapleton, H. (2010). Bonny babies? Motherhood and nurturing in the age of obesity. *Health, Risk and Society, 12*(4), 369–383.

La Fontaine, J. S. (1998). *Speak of the devil: Tales of satanic abuse in contemporary England.* Cambridge/New York: Cambridge University Press.

Laming, L. (2003, January). *The Victoria Climbié inquiry: Report.* London: HMSO. Available at: http://webarchive.nationalarchives.gov. uk/20130401151715/http://www.education.gov.uk/publications/eOr- deringDownload/CM-5730PDF.pdf. Accessed 11 Dec 2015.

Lasch, C. (1977). *Haven in a heartless world: The family besieged.* New York/ London: Basic Books.

Lasch, C. (1979). *The culture of narcissism: American life in an age of diminishing expectations.* New York/London: WW Norton and Company.

Lasch, C. (1984). *The minimal self: Psychic survival in troubled times.* New York/ London: WW Norton and Company.

Lee, E., Bristow, J., Faircloth, C., & Macvarish, J. (2014). *Parenting culture studies.* Basingstoke: Palgrave Macmillan.

Lyons, R. (2011). *Panic on a plate: How society developed an eating disorder.* Exeter: Imprint Academic.

Mannheim, K. (1952). *Essays on the sociology of knowledge.* Paul Kecskemeti (Ed.). London: Routledge & Kegan Paul Ltd.

May, T. (2015, March 14). Child abuse in the UK runs far deeper than you know. *Daily Telegraph.* Available at: http://www.telegraph.co.uk/news/uknews/ crime/11471282/Theresa-May-Child-abuse-in-the-UK-runs-far-deeper- than-you-know.html. Accessed 9 Dec 2015.

Mays, J. B. (1961). Teen-age culture in contemporary Britain and Europe. *Annals of the American Academy of Political and Social Science, 338*, 22–32.

McAlinden, A.-M. (2010). Vetting sexual offenders: State over-extension, the punishment deficit and the failure to manage risk. *Social Legal Studies, 19*(1), 25–48.

Munro, E. (2003, September 10). This would not have saved Victoria. *Guardian*.
Munro, E. (2007). *Child protection*. Thousand Oaks: Sage.
Murray, S. B. (1996). "We all love Charles": Men in child care and the social construction of gender. *Gender & Society, 10*(4), 368–385.
Murray, S. B. (2001). When a scratch becomes 'a scary story': The social construction of micro panics in center-based child care. *The Sociological Review, 49*(4), 512–529.
NHS. (2015). Change 4 Life. Available at: http://www.nhs.uk/change4life/Pages/change-for-life.aspx. Accessed 9 Dec 2015.
Office for National Statistics. (2014, September 23). Intergenerational transmission of disadvantage in the UK & EU. Available at: http://www.ons.gov.uk/ons/rel/household-income/intergenerational-transmission-of-poverty-in-the-uk---eu/2014/blank.html. Accessed 18 Dec 2015.
Palmer, S. (2007). *Detoxing childhood: What parents need to know to raise happy successful children*. London: Orion.
Parton, N. (2006). *Safeguarding childhood: Early intervention and surveillance in a late modern society*. Basingstoke: Palgrave Macmillan.
Phillipson, C. (2013). *Ageing*. Cambridge: Polity Press.
Pilcher, J. (1995). *Age and generation in modern Britain*. Oxford: Oxford University Press.
Piper, H., & Stronach, I. (2008). *Don't touch! The educational story of a panic*. London/New York: Routledge.
Piper, H., Powell, J., & Smith, H. (2006). Parents, professionals, and paranoia: The touching of children in a culture of fear. *Journal of Social Work, 6*(2), 151–167.
Postman, N. (1994). *The disappearance of childhood*. New York: Vintage.
Randall, V. (2000). *The politics of child daycare in Britain*. Oxford: Oxford University Press.
Riley, D. (1983). *War in the nursery: Theories of the child and mother*. London: Little, Brown Book Group Limited.
Schoon, I., Barnes, M., Brown, V., Parsons, S., Ross, A., & Vignoles, A. (2012, September 27). *Intergenerational transmission of worklessness: Evidence from the Millennium Cohort and the longitudinal study of young people in England*. Institute of Education & National Centre for Social Research; Department for Education. Available at: https://www.gov.uk/government/uploads/system/uploads/attachment_data/file/183328/DFE-RR234.pdf. Accessed 18 Dec 2015.
Sikes, P., & Piper, H. (2010). *Researching sex and lies in the classroom: Allegations of sexual misconduct in schools*. London/New York: Routledge.
Skenazy, L. (2010). *Free-range kids: How to raise safe, self-reliant children (without going nuts with worry)*. San Francisco: Jossey-Bass.
Somerville, J. (2000). *Feminism and the family: Politics and society in the UK and USA*. Hampshire/London: Macmillan Press.

Spencer, B., Cohen, T., & McTague, T. (2015, October 5). CBeebies called on to help the government tackle child obesity by saying "chips are bad" on TV. *Daily Mail.* Available at: http://www.dailymail.co.uk/news/article-3260782/CBeebies-ordered-help-government-tackle-child-obesity-saying-chips-bad-TV.html. Accessed 9 Dec 2015.

Stanfors, M., & Scott, K. (2013). Intergenerational transmission of young motherhood. Evidence from Sweden, 1986–2009. *The History of the Family, 18*(2), 187–208.

Ward, S., & Eden, C. (2009). *Key issues in education policy.* Thousand Oaks: Sage.

Webster, R. (2005). *The secret of Bryn Estyn: The making of a modern witch-hunt.* Oxford: Orwell Press.

CHAPTER 5

Gender and the Intimate Politics of Reproduction

Abstract Policy interest in the problem of generations has for a long time had a naturalistic quality, expressed in a preoccupation with demographic trends, and the ideology of eugenics. It has also presumed an interest in the domain of social reproduction, situating the family as a cause of, and solution to, social problems. This chapter explores the way that changes in women's social position over the twentieth century have both allowed women to participate fully as members of 'social generations', and opened up the sphere of reproduction to intensified scrutiny and management. Relations within, and between, generations are conceptualised in increasingly brittle terms.

Keywords Demography • Suffragettes • David Willetts • Welfare state • The Sixties • Eugenics

Questions of how children are taught in schools, and how they are raised in their homes, have in recent years become central to the policy agenda. One consequence, as previous chapters have noted, has been an increasing bureaucratisation of relations between the generations, expressed in the development of policies around parenting, safeguarding, and the micromanagement of teaching practice. These developments reflect the orientation of social policy around questions of reproduction. By this, we mean both the biological reality of human reproduction (the fact that there is 'one born every minute') and social reproduction: the 'combination of familial, economic, state, and local community structures' that 'shape

J. Bristow, *The Sociology of Generations*,
DOI 10.1057/978-1-137-60136-0_5

how, where, and with whom children are raised or the end of life is negotiated' (Cole and Durham 2007, p. 6).

To some degree, it is possible to argue that the current focus on 'generation' merely follows the trajectory of the development of social policy. The image of the 'Victorian family' captures the sense in which the nineteenth-century age of ideology regarded both the biological and social aspects of reproduction as properly situated in the private, or domestic, sphere of life. The private sphere was largely the responsibility of women, while the public domain of politics, policy, ideas, and culture was the preserve of men. But by the early twentieth century, the stark divide between the public and private spheres was beginning to break down. The state came to be seen as increasingly responsible for the welfare and well-being of its citizens, and both biological and social reproduction came to the fore as sites for policy intervention.

Thus, as we discuss below, attempts to find solutions to social problems such as poverty and ill health by focusing on the biology of reproduction have a long and ignoble history, epitomised by eugenic policies designed to limit the number of children born to poor women and increase the number born to middle-class women. Eugenic ideology developed alongside a contrasting, and equally influential, policy perspective that perceived social problems as the product of wider tensions in the realm of economic relations and social structures, and emphasised that the state should play a proactive role in enhancing welfare, through the promotion of education, healthcare, and housing. Despite their clear differences, however, both perspectives rested on the assumption that the structures and relationships of reproduction held the key to the resolution of wider social problems.

In this regard, the tension between the public and private realms has framed the problem of generations for over a century. What is striking about the present period, however, is the extent to which the discourse of 'generationalism' seems to view *no distinction* between the public and private realms of life. As the core institutions of the welfare state—publicly funded schools, hospitals, pensions, housing, and benefits—are reconfigured, there is a rhetorical attempt to 'return' the responsibility for raising children and caring for the elderly to the realm of the family. But the family that is conceptualised by policy today is not a private institution that is separate from the state: it is cast as a 'partnership' with the state, in which parents are positioned as the intermediaries for expert advice and government prescription (Bristow 2010). Just as the role of the teacher has been gradually redefined, from that of 'an individual professional trusted to make judgements about the needs of the community and individuals' to 'a technician employed to carry out national government policy' (Ward and Eden 2009, p. 102), so parents have been re-cast as caregivers, directly accountable for their 'competence' in raising their children according to the orthodoxy of the time (Furedi 2001; Gillies 2011; Lee et al. 2014).

The trends described in the previous chapters, towards the de-professionalisation of teaching and the bureaucratisation of child protection, have their roots in a wider anxiety about the capacity of older generations to equip young people for life in the present day. Yet they seek their resolution in the realm of intimate relationships, seeking to limit the influence of parents, teachers, and the wider adult community upon the children whom they are rearing and educating. In this respect, the age-old problem of generations is collapsed together with present-day problems confronting politics and policy.

DEMOGRAPHY AND THE 'NATURAL EXISTENCE' OF GENERATIONS

'Like the term "reproduction", with its nod to women's fertility, productive labor, and the mechanical duplication of a well-designed product, the root of our term, "generation", covers a wide range of conceptual territory,' write Cole and Durham (2007, p. 15). As we have noted, 'generation' denotes the biological reality of being, the historical reality of living, and the epistemological problem of knowing. This can make it a slippery concept. When policymakers—or anyone else—talk about 'a generation', it is often unclear whether they are talking about a cohort of people roughly the same age, or a historical period, or a particular outlook on life—or some combination of all three.

The multiple meanings attached to the concept of 'generation' have made it an attractive term for those seeking a version of a grand narrative at a time when the political ideologies of the past have been exhausted. Yet it is also why the concept of generation is problematic when it is mobilised for the purpose of political claims-making.

One example of the tensions that emerge with the politics of generationalism is provided by the ways in which the generation known as the 'Baby Boomers' has been constructed as a social problem in Britain and in the USA. As I have discussed elsewhere (Bristow 2015), the claim that the size of the Baby Boomer generation is the cause of myriad social problems has become a powerful narrative in recent years. This claim has both quantitative and qualitative dimensions, relating both to the relative size of the cohort born in the twenty years after the Second World War and the attitudes and behaviours attributed to the 'Sixties generation'.

The Boomers, as a large generation that is now reaching retirement age, are blamed for the unsustainable cost of pensions, health and social care, and the shortage of housing stock, meaning that younger people are now struggling to be able to afford to buy their own home. This focus, on the absolute and relative size of the Boomer cohort, appears to present the problem of generations as primarily one of biological reproduction: there were, allegedly, too many babies born between 1945 and 1965 and too few born in subsequent years (Preston 1984). In this respect, the problem of the Baby Boomers appears as a manifestation of a preoccupation with demography that has underpinned discussions about the problem of generations for many years.

Of course, when designing social policies, the insights offered by demography are crucial. From the amount of public money needed to fund state pensions, to the capacity of the health service, to the number of primary school places that will be needed in four years' time, it is essential to know both how many people there are, and what other trends—including ill health, death rates, and migration—should be taken into account in order to ensure that services can be properly developed. Furthermore, in the context of the sociology of generations, the work of demographers such as MacInnes and Díaz (2009), Lesthaeghe (2010), and Davis (1940), which has attempted to integrate trends in birth and death rates with an understanding of their wider social and cultural context, has yielded some important insights into the way that ideas about children and the family have developed in the post-war period. For example, Falkingham's (1997) 'demographic profile' of the Baby Boomers challenges some of the claims made about the size and influence of this cohort and points out some important international differences in the character of the 'Baby Boom' that tend to be missed in the use of this label.

But the demographic consciousness that informs the current policy debate does not signify an appreciation of the signficance of population trends. It is a particular way of thinking, which can distort attempts to understand the implications of demographic changes through assuming that social changes are driven primarily by numbers, and by using demographic facts as 'evidence' of social or cultural problems that stem from quite different sources.

Thus, my study of the cultural script of the 'Baby Boomer problem' found that claims-makers on the conservative end of the political spectrum tended to blame the 'Sixties generation' for the crisis of traditional values and high levels of public spending, while those on the liberal left blamed them for their failure to preserve the welfare state and the availability of social housing. Both versions of this critique might take the form of a demographic analysis about the size of the Boomer cohort, but they are underpinned by a reaction against wider social, cultural, and economic changes. Claims about the challenges caused by the size of the Boomer cohort have a moralised character, bound up with wider claims-making about the 'selfish', irresponsible, and reckless attitudes associated with the 'Sixties generation' (Bristow 2015).

Many of the specific features of 'Boomer Blaming' have been challenged, both ideologically and empirically. Research indicates that there are wide variations in the health, wealth, and levels of dependency of those individuals captured within definitions of the Baby Boomer cohort, and this complicates the claim that the Boomers have, in general, experienced untrammelled good fortune and used more than their 'fair share' of public resources. Meanwhile, critics have suggested that the narrative of 'generationalism' (White 2013) is used to deflect unease about the restructuring of the welfare state away from policymakers, by presenting it as an unavoidable response to the size and behaviour of a particular, ageing cohort (Walker 1996). The effect of this is to mobilise a sentiment of intergenerational conflict and injustice, in order to provide a warrant for what the recently established lobby group the Intergenerational Foundation calls 'a programme of "intergenerational rebalancing"', which seeks to transfer power and resources from the older generation, purportedly for the benefit of their young (Intergenerational Foundation 2015).

Generationalism is not merely, however, a discussion about resources. In using the term to denote 'the systematic appeal to the concept of generation in narrating the social and political', White (2013, p. 216) recognises that generationalism is, fundamentally, about a particular way of thinking and knowing. We suggest that influence of 'generationalism' in this respect is not due to changes in the relative size of generations, nor to a decline in the quality of relations between generations, as claims-makers would have it. Rather, it is due to a shift in our knowledge and understanding of social problems, which situates the primary cause and solution to social problems within the realm of reproduction.

GENDER AND THE GENERATIONAL CONTRACT

In his influential book *The Pinch: How the Baby Boomers Took Their Children's Future—and Why They Should Give It Back,* David Willetts (2010), then Minister of State for Universities and Science, insists that the central problem with British social policy today lies in its failure to attach 'sufficient value to the claims of future generations'. His argument is premised on a particular diagnosis of the problem of the 'Baby Boomer' generation, which, he claims, has monopolised economic, social, and cultural resources, and thereby 'weakened many of the ties between the generations' (Willetts 2010, p. 260).

Willetts presents the 'Baby Boomer' problem as emblematic of 'an intellectual failure: we have not got a clear way of thinking about the rights of future generations. We are allowing one very big generation to break the inter-generational contract because we do not fully understand it' (Willetts 2010, pp. 260–1). He insists, 'That is where politics comes in.' Thus, he argues, politics should be oriented more around relations between the generations, with policy playing the mediating role between the parent generation and their children.

The overt focus on generations in present-day policy clearly expresses the tension between past, present, and future described in previous chapters. When public policy comes to focus on the intricacies and intimacies of relations between the generations, presupposing that such relations should be re-ordered around the presumed demands of the present or a particular, already-imagined future, this indicates an important shift in the social and political imagination, which sees the adult world as having a destructive impact on the younger generation, rather than one that is constructive and protective. 'The UK, like other developed economies, has engaged in fiscal, educational, health and environmental child abuse,' claims Laurence Kotlikoff, professor of economics at Boston Massachusetts Institute of Technology and former World Bank economist (Intergenerational Foundation 2015).

Concerns about the 'generational contract' may have their origins in conservative thought—but they are shared by risk society theorists and others concerned about the perceived disintegration of the family and the welfare state. In her 2002 book *Reinventing the Family: In Search of New Lifestyles* Elisabeth Beck-Gernsheim dedicates a chapter to 'Generational Contract and Gender Relations'. She begins by noting that during the women's movement of the late 1960s and early 1970s, 'relations between men and women, including not least relations of misunderstanding and dependence, were no longer just the material of private conversation or conflict but excited the attention of the media, politicians and public opinion' (Beck-Gernsheim 2002, p. 64).

During the 1980s, the question of relations *between* the generations also 'began to detach itself from the horizon of the merely private'. Against a backdrop of anxiety about falling birth rates and ageing populations, and the sustainability of pensions and nursing care, '[t]he relationship between old and young people, both quantitative and qualitative, became a topic for political commissions, academic studies and population forecasts' (Beck-Gernsheim 2002, pp. 64–5). In setting out her intention to inquire 'what the relationship between the generations and the relationship between the sexes have to do with each other', Beck-Gernsheim poses the question:

> Is it a historical accident that both are no longer treated as a matter of course, that both are sure to cause agitation among the public? Or are the two associated with and dependent upon each other in a number of ways? (Beck-Gernsheim 2002, p. 65)

The changing position of women in society, and the concomitant shifts in the relations between the sexes, is a highly significant development that has informed the present-day policy focus on reproduction. Below, we outline some of the trends that have coalesced to bring the intimate relations between the generations, and between the sexes, into the frame of policy regulation.

THE BIRTH OF SOCIAL POLICY AND THE PROBLEM OF REPRODUCTION

As we noted in the introductory chapter, when Mannheim formulated the sociology of generations in the 1920s, the story of generations was primarily about, and written by, men. Precisely because women's role was largely focused on reproduction, and confined to the private sphere, women were not yet fully part of the story of generations.

Vera Brittain's memoir *Testament of Youth* evokes the position of educated, middle-class women around the time of the First World War. Over several hundred pages, Brittain (1984 [1933]) describes the experience of feeling compelled to participate in the war effort, yet confined to a role as a nurse, and watching her brother, fiancé, and male friends perish. *Testament of*

Youth is ripe with references to 'our generation', and as such can be seen to be as much a product of the 'Generation of 1914' as the writings of Rupert Brooke, Wilfred Owen, or Siegfried Sassoon (see discussion in Badenhausen 2003; Schwarz 2001). Yet it is one woman's engagement in a drama dominated by men.

However, women were already being brought into the wider story of generations. *Testament of Youth* also gives us vignettes of the moment when university degrees were first conferred on women, and when women gained the right to vote. Right up to the beginning of the war, the campaigns of the militant suffragettes had challenged both women's confinement to the domestic sphere and the cultural trappings that legitimised this. As Susan Kent (1987) argues, in her account of *Sex and Suffrage in Britain, 1860–1914:*

> Nineteenth-century feminists argued... that the public and the private were not distinct spheres but were inseparable from one another; the public was private, the personal was political. Suffragists perceived their campaign as the best way to end a 'sex war' brought about by separate sphere ideology—an ideology that finally reduced women's identity to a sexual one, encouraged the view of women as sexual objects, and perpetrated women's powerlessness in both spheres. (p. 5)

It is questionable whether the notion that 'the personal is political'—a slogan that, as we have noted, has its roots in the political reconfigurations of the 1960s—should be associated with the movement for women's suffrage in the late nineteenth and early twentieth centuries. But the identification of 'separate sphere ideology' as a core factor in the denial of equal rights to women is well documented; and the campaign waged by the Suffragettes, along with the experience of the First World War, centrally attacked naturalised cultural assumptions about male and female difference, expressed in the double standards applied to sexual behaviour and morality. 'Why is it that men's blood-shedding militancy is applauded and women's symbolic militancy punished with a prison cell and the forcible feeding horror?' wrote Emmeline Pankhurst in her impassioned autobiography. She answered:

> It means simply this, that men's double standard of sex morals, whereby the victims of their lust are counted as outcasts, while the men themselves escape all social censure, really applies to morals in all departments of life.

Men make the moral code and they expect women to accept it. They have decided that it is entirely right and proper for men to fight for their liberties and their rights, but that it is not right and proper for women to fight for theirs. (Pankhurst 2015 [1914], loc. 2873)

Challenging sexual stereotypes and double standards was an important feature of the campaign for women's equality in the early part of the twentieth century. However, the gradual diminishing of the power of separate sphere ideology was also made possible by the nascent welfare state. Millicent Fawcett observed in 1886 that 'women's suffrage will not come... as an isolated phenomenon'. Rather, she said, 'it will come as a necessary corollary of the other changes which have been gradually and steadily modifying, this century, the social history of our country' (cited in Pugh 2000, p. 63). When women finally gained the right to vote on equal terms to men in 1928, this was in a context where public life was already becoming organised around the principles of welfare and social administration.

From the early twentieth century, the government had begun to take on board functions that had previously been seen as restricted to the private sphere. '[A]lthough the Victorians regarded social welfare as the proper preserve of local authorities, by the 1890s pressure had grown for a national government to assume a much larger measure of responsibility,' writes Pugh. He continues:

As a result, the post-1906 Liberal government intervened over free school meals, school medical services, old age pensions, health and unemployment insurance, labour exchanges, and minimum wages. In the process they went a long way to turning national politics into local politics and, as a crucial if unintended by-product, they materially diminished the distinction between the male and female spheres. (Pugh 2000, p. 76)

In this regard, 'social reform became the bridge between local and national politics across which women could advance without posing a fundamental threat to conventional thinking about gender' (Pugh 2000, p. 136).

'The fact that the modern age emancipated the working classes and the women at nearly the same historical moment must certainly be counted among the characteristics of an age which no longer believes that bodily functions and material concerns should be hidden,' argues Arendt (1998 [1958], p. 73). Problems such as poverty, illness, and education, which had previously been 'hidden' within the family, were now seen as issues for the state to take an interest in and make public policy about; and the sphere of reproduction became subject to professional scrutiny and intervention (Lewis 1980).

In British social policy, this took two main forms. For the Fabian Society, formed in 1884 and led by Sidney and Beatrice Webb, 'solutions to the problem of poverty lay in the correct application of professional knowledge appropriate to the cause of destitution', thereby informing 'the provision of state services appropriate to cure poverty: medical care for the sick, residential care for the old, work for the unemployed, disciplinary camps for the "workshy," and so on' (Whiteside 2012, p. 119). These, and related, developments were eventually institutionalised in the postwar welfare state, which positioned social policy as the way to ameliorate the 'five giant social evils' famously identified by William Beveridge: ignorance, idleness, disease, squalor, and want.

On the other hand, for the eugenics movement, led by Galton and Pearson, 'inheritance explained the larger part of poverty through the perpetuation of mental, moral, and physical weakness'. In consequence:

> They demanded that 'unfit' paupers be sterilized and that official inspection be required prior to marriage before the happy couple be permitted to wed and breed. Such ideas, later elaborated in Hitler's Germany, carried heavy racist overtones and appealed to imperial sentiment through references to Darwinian theories concerning 'the survival of the fittest'. (Whiteside 2012, p. 119)

Because eugenic ideology cast biological reproduction as both the cause and the solution to social problems, women in particular became the focus of these ideas. For example, debates about the need to manage the pain of childbirth, either through the provision of obstetric anaesthesia and analgesia or through the promotion of 'natural birth' techniques that encouraged women to experience the pain of birth as a spiritually transcendental experience, often emphasised the need to counter falling birth rates among middle-class women so as to improve the moral and physical health of the new generation (Moscucci 2003).

The development of both the technology of modern family planning and the idea that family size and timing could, and should, be controlled by individual women was closely linked both to the ideology of eugenics and the feminist rhetoric of women's emancipation. This was personified by Margaret Sanger in the USA and by Marie Stopes in Britain. Sanger founded the American birth control movement and, later, the Planned Parenthood Federation of America. In doing so, she pushed forward the case for women to have control over their bodies: an important principle of women's freedom (Planned Parenthood Federation of America 2004). At the same time, many of her ideas were born out of a desire to improve the 'quality of the race' (Sanger 1921) through control over reproduction. For example, an article by Sanger titled 'The Eugenic Value of Birth Control Propaganda', published in 1921, argued:

> The doctrine of Birth Control is now passing through the stage of ridicule, prejudice and misunderstanding. A few years ago this new weapon of civilization and freedom was condemned as immoral, destructive, obscene. Gradually the criticisms are lessening—understanding is taking the place of misunderstanding. The eugenic and civilizational value of Birth Control is becoming apparent to the enlightened and the intelligent. (Sanger 1921)

Sanger went on to 'touch upon some of the fundamental convictions that form the basis of our Birth Control propaganda', which, she argued, 'indicate that the campaign for Birth Control is not merely of eugenic value, but is practically identical in ideal, with the final aims of Eugenics'. She outlined these as follows:

> First: we are convinced that racial regeneration like individual regeneration, must come 'from within'. That is, it must be autonomous, self-directive, and not imposed from without. In other words, every potential parent, and especially every potential mother, must be brought to an acute realization of the primary and central importance of bringing children into this world.
>
> Secondly: Not until the parents of the world are thus given control over their reproductive faculties will it ever be possible not alone to improve the quality of the generations of the future, but even to maintain civilization even at its present level. Only by self-control of this type, only by intelligent mastery of the procreative powers can the great mass of humanity be awakened to the great responsibility of parenthood.
>
> Thirdly: we have come to the conclusion, based on widespread investigation and experience, that this education for parenthood and of parenthood must be based upon the needs and demands of the people themselves. An idealistic code of sexual ethics, imposed from above, a set of rules devised by high-minded theorists who fail to take into account the living conditions and desires of the submerged masses, can never be of the slightest value in effecting any changes in the mores of the people. Such systems have in the

past revealed their woeful inability to prevent the sexual and racial chaos into which the world has today drifted. (Sanger 1921)

Sanger's aim was to 'effect the salvation of the generations of the future—nay of the generations of today' by building on the interest in birth control. 'In answering the needs of these thousands upon thousands of submerged mothers,' she wrote, 'it is possible to use this interest as the foundation for education in prophylaxis, sexual hygiene, and infant welfare. The potential mother is to be shown that maternity need not be slavery but the most effective avenue toward self-development and self-realization. Upon this basis only may we improve the quality of the race' (Sanger 1921).

In Britain in 1921, Marie Stopes established the first 'family planning clinic' in Islington, North London, where women were fitted with cervical caps. Stopes's aim was to enable women to avoid having children that they did not want or could not support and to reduce the incidence of abortion by the prevention of unwanted conceptions. 'Stopes's multifaceted campaign was fraught with contradiction,' writes Neushul (1998):

> On the one hand she was an ardent eugenicist who once commented that a third of the men in England should be sterilized, 'starting with the ugly and unfit', while on the other hand she was a concerned clinician who invariably placed an individual's happiness over eugenic ideals. (Neushul 1998, p. 246)

Stopes was deeply influenced by neo-Malthusian ideas, which saw poverty as a problem of population numbers, and by the wider eugenic outlook, which held that society could be improved by manipulation of heredity: hence her establishment of the 'Society for Constructive Birth Control and Racial Progress'. Yet she also saw contraception as the basis of 'marital bliss' as it allowed for the full expression of women's passions and pleasure and was concerned that women's lives were diminished by too much childbearing.

Cohen (1993) suggests, in her study of 'Private Lives in Public Spaces: Marie Stopes, the Mothers' Clinics and the Practice of Contraception', that historians have presented a 'remarkably lopsided account of the birth control movement' in focusing on its eugenic ideas, which fails to account for the extent to which Stopes, 'in constructing a model of medical intervention designed to enable women to use contraception... subordinated eugenic and political considerations to her overriding concern for the individual woman's health and happiness' (Cohen 1993, p. 97). And indeed, as Marks's (2010) study of the history of the contraceptive pill in Britain and America shows, this tension between political agendas of eugenics and population control, and a commitment to meeting women's genuine desire and need for a way of managing their own fertility, would continue to frame the debate around birth control for the rest of the twentieth century.

The ideology of eugenics was discredited by the horror of National Socialism and the Holocaust; and in the immediate post-war period, policies tended to focus on presumed structural, environmental, and cultural causes of social problems. However, attempts to intervene in the biology of reproduction continued, though the language of eugenics was tempered. For example, a key motivation for the passage of the 1967 Abortion Act, which allowed abortion in circumstances where two doctors agreed that a pregnancy would pose a greater risk to a woman's health than would a termination, was the belief that too many women were having children that they did not want and could not afford or cope with. In education, Chitty (2013) has suggested that '[a] number of the terms and phrases we now use in relation to debates about human intelligence, and the role that can be ascribed to it in fostering the development of the human race, can, in fact, be attributed to (Sir) Francis Galton' (Chitty 2013, p. 352).

As these examples indicate, by the 1960s, the ideology and institutions of the welfare state were already engaged with the realm of reproduction, seeing this as a realm that needed to be monitored and managed. However, policy engagement with the problem of generations via the central institutions of the family and the school remained at a distance. The family was considered to be the institution primarily responsible for the socialisation and welfare of children, and family policy took an 'implicit' form, focused on developments that would affect the family, but which lay outside it: for example, compulsory education, school meals, childhood immunisation programmes, and the provision of universal healthcare. Education was provided, funded, and monitored by the state and was the subject of countless official inquiries and reports (see Gillard 2011). Yet, as previous chapters have indicated, schools and teachers held a considerable degree of professional autonomy in relation to curriculum content and teaching practice.

The social and cultural upheavals of the 1960s, however, would cause a shift in the balance that was perceived to lie between the remit of policy and the privacy of the intimate realm.

THE SIXTIES AND THE 'WAR OVER THE FAMILY'

'By the mid-1960s, as a cultural revolution was spreading through Western societies like wildfire, there became a great debunking of what the anthropologist R. Birtwhistell called the "sentimental myth" of the family,' write Berger and Berger (1983) in *The War over the Family*. They continue:

> Various political and ideological trends came together in this demolition job. The New Left was interested in exalting the public sphere over all forms of private life, in having women return to the sphere of work, in collectivizing child-rearing and in 'non-repressive sexuality'. The feminists, wanting to liberate women from domesticity, were against the privatized and child-centred 'bourgeois family'. Populationists were against the 'frightful reproductive potential' of the family. And the professional complexes, growing in power (mainly by government support) and legitimated in their role by a functional definition of the family, were obviously interested in proclaiming the inefficiency of the family in providing 'services' that they had a claim to. (Berger and Berger 1983, pp. 62–3)

This summary of developments highlights that what was important about the 1960s was the change in *ideas* about the family, rather than sudden changes in the form or function of the family itself. This insight remains true in the present day. The relative normalisation of cohabitation, divorce, homosexual parenthood, and lone parenthood—the changes that are generally focused on in debates about the family—are interesting and significant and have certainly affected the appearance of the modern family. But from a sociology of knowledge perspective, our interest is in the way that the qualities of the family, as a set of relations for having and rearing children, are perceived.

By the time of the Culture Wars of the 1980s, Berger and Berger discerned 'three major alignments on family issues in America'. The first was 'the radical-to-reformist coalition rooted in the movements of the 1960s, moderated somewhat by middle age and by the compromises of political horse-trading but still marching under the old banners of liberation'—this is the camp that is widely seen to be personified today by the 1960s generation of Baby Boomers. The second was 'the new "pro-family" camp, more brash and uncompromising by its very youth, marching in step with the general veering towards conservatism in the national mood'—an align-ment forged in reaction to the changes of the 1960s, largely defined by its opposition to abortion, pornography, and promiscuity and its defence of marriage (Berger and Berger 1983, pp. 14–5).

The third alignment identified by Berger and Berger was 'the combi-nation of professionals, academics and bureaucrats who make the family their field of expertise, advocacy, and management'—a development of the social policy thinking and interventions described above in this chap-ter, whose role in the 'safeguarding' agenda is discussed in Chap. 4. Berger and Berger note that for all three camps, the family has become a central political battleground:

> All these people share the very formal proposition that the family has 'prob-lems' or even *is* a 'problem', but, of course, they vary sharply in their under-standing of what it is that is problematic and what should be done about it. Indeed, the 'solutions' of the first camp are an essential part of what the second camp sees as a 'problem', and vice versa, while the family experts often have 'problems' that nobody else on the scene is capable of perceiving. (Berger and Berger 1983, pp. 14–5)

In this respect, the 'war over the family' has resulted in the exposure of intimate relationships—between men and women and between parent and child—to political scrutiny and policy-wonkery. Family policy has assumed a more 'explicit' form, seeking to intervene directly in relations between parents and children and to achieve particular parenting 'out-comes' (Daly 2010; Lee et al. 2014; Lewis 2011). Generation becomes central to what Cole and Durham term 'the intimate politics of glo-balisation': the notion that 'if we want to understand the contemporary moment, then we must focus not only on broad questions of labor, capi-tal, and the management of populations, or the movement of commodi-ties' but that we must also 'take seriously the fact that these phenomena take place in the context of familial and generational relations' (Cole and Durham 2007, p. 17). Yet the very concept of an 'intimate politics' is revealing, for the way it conceptualises relations between the generations in the present day.

GENERATIONS AND THE TRANSFORMATION OF INTIMACY

In *The Normal Chaos of Love* (1995), Beck and Beck-Gernsheim examine the nature of love in the context of changing social structures and moral norms. In an age of women's equality, they argue, the solidity of marriage and the family as a basis for organising intimate relationships has dissolved, ushering in both more freedom and 'democracy' in the sphere of private life and the greater potential for chaos.[1]

Beck and Beck-Gernsheim explain that, as women are no longer dependent upon men in terms of financial need or social status, there is less incentive to remain committed to an unhappy relationship. This, in turn, invites more tensions. An autonomous woman demands more from a free relationship than from a relationship in which she is subordinate; a man, stripped of his traditional breadwinner role, is less certain of his identity and more prone to questioning his own role. Intimate relationships thus become fraught and temporary: so far as the family goes, 'the child becomes the last remaining, irrevocable, unique primary love object. Partners come or go, but the child stays' (Beck and Beck-Gernsheim 1995, p. 37).

The Normal Chaos of Love contains a great deal of insight in the extent to which it accepts the magnitude of changes that intimate relationships have undergone in the past quarter-century and the impact that this has upon the stability of the nuclear family. Women's increased independence and autonomy in relation to the family is linked with the shifts in morality away from condemnations of single motherhood, divorce, or same-sex relationships—a process that is generally accepted as a good thing, allowing individuals more freedom in the kind of intimate relationships they have and the way that they choose to conduct them. And, of course, greater freedom and choice bring new tensions into play. As Beck and Beck-Gernsheim explain:

> It is no longer enough to just get along with each other. People want more, they are in search of 'happiness and fulfilment', the American dream, 'the pursuit of happiness' in their own little home. (Beck and Beck-Gernsheim 1995, p. 93)

However, there is an awkward disjuncture between these authors' description of the changing situation and the bleak and brooding conclusions drawn along the way. The basis of their analysis is that love relationships have become less inherently stable, more difficult to negotiate, and subject to higher expectations—and that this therefore causes a host of major problems. But why should people's pursuit of happiness in their intimate relationships necessarily mean that 'disappointments are inevitable' and '[f]urthermore, the dream turns into a trap, arousing hopes which cannot be satisfied'? We might also wonder why, following a thoughtful analysis of the increasingly emotionally charged character of child-rearing and the role of the child in a modern family relationship, we are presented with the following blistering conclusion:

> Love is one of our great achievements, the foundation of our relationships between men and women, parents and children—but we cannot have it without its darker sides, which sometimes emerge for a second and sometimes linger for years: disappointment, bitterness, rejection and hatred. The road from heaven to hell is much shorter than most people think. (Beck and Beck-Gernsheim 1995, p. 139)

This is more than the truism that there is a thin line between love and hate. Rather, Beck and Beck-Gernsheim are warning that the new freedoms we have about who and how we love today are, in fact, deeply dangerous. Without the limits imposed upon passion by the traditional institutions and conventions of the family, which implicitly prevented people from expecting too much or complaining too much, people now have the capacity to give free reign to their passions and emotions, with worrying consequences: 'The more intense our feelings are, the more likely we are to suffer from them, from the mistakes, misunderstandings and complications they bring about' (Beck and Beck-Gernsheim 1995, p. 100).

According to this perspective, the problem with love is not what has changed, but what has stayed the same. For all that the institutional upheavals of marriage and family life are focused upon as a cause of individuals' greater insecurity and inability to sustain a love relationship, ultimately the problem is presumed to be passion, or 'intensity of feeling', itself.

The theme of how individuals manage their intimate relationships in an increasingly fragmented and insecure society forms the basis of sociologist Anthony Giddens's conceptualisation of 'plastic sexuality' and the 'pure relationship', in his influential theory of the 'transformation of intimacy' (Giddens 1992). 'Plastic sexuality', Giddens argues, 'is decentred sexuality, freed from the needs of reproduction ... Plastic sexuality can be moulded as a trait of personality and thus is intrinsically bound up with the self' (Giddens 1992, p. 2). The 'pure relationship', meanwhile, 'refers to a situation where a social relation is entered into for its own sake, for what can be derived by each person from a sustained association with another; and which is continued only in so far as it is thought by both parties to deliver enough satisfaction for each individual to stay within it' (Giddens 1992, p. 58).

At a descriptive level, both these concepts appear as useful summaries of how intimate relationships have changed as a consequence of reproductive freedom and social progress. The separation of sex from reproduction, made possible by the widespread availability and acceptability of contraception and abortion, has indeed transformed the basis upon which intimate relationships develop or wane. Heterosexual relationships can be taken 'all the way' and retreated from without the once-dominant concern about an accidental pregnancy and attendant responsibility towards a child; homosexual relationships, once viewed by mainstream society as pathological and subversive, are now accepted in much of the Western world as just another lifestyle choice and legitimised through marriage.

The magnitude of these changes, both at the level of sexual behaviour and attitudes towards sexual behaviour, is confirmed by the recently published Third National Survey of Sexual Attitudes and Lifestyles (Natsal-3). Lead authors Kaye Wellings, of the London School of Hygiene and Tropical Medicine, and Anne Johnson, of University College London, explain that today, 'sexual activity is not primarily, or even necessarily, about reproduction'. They continue:

> In a growing number of contexts globally, the separation of sexual activity from reproduction is well under way as contraception, abortion, and assisted reproduction have weakened the natural link. Sexual behaviours that are not essential to conception have become easier to discuss and have gained greater acceptance; they include masturbation, oral and anal sex, same-sex practices, and sex in groups among whom reproduction may not be possible or might have conventionally been deemed inappropriate. In many cultural contexts, what was once seen as deviance or perversion is increasingly referred to as diversity. (Wellings and Johnson 2013, pp. 1760–1)

However, Giddens insists that the 'pure relationship' concept should not be taken at face value, for: 'A pure relationship has nothing to do with sexual purity, and is a limiting concept rather than only a descriptive one' (Giddens 1992, p. 58). It is not that individuals are freed from the shackles of social convention in order to engage in intimate relationships based on their emotional feelings for one another. Rather, the significance of intimate relationships is transformed, from being framed in terms of two people's relationship with one another to being framed in terms of how an individual feels about him- or herself.

For Giddens, the transformation of intimacy means a slow rejection of romantic love in favour of the careful management of emotion—part of a calculated strategy of identity creation, mediated not through the principles of passion but through therapeutic self-regulation. The purpose of a relationship is not that the individual becomes subsumed into some other bigger, grander unit—the couple, the family—but that they learn from the experience of his emotional attachment in the course of their own self-fulfilment. The presentation, here, is of intimate relationships that have been stripped to their most instrumental core. In the era of the 'pure relationship', conducted against the backdrop of risk consciousness, relationships are framed according to whether they are 'good for you' or 'bad for you'—in other words, in isolation from their wider generational context. At a cultural level, this denudes intimacy of its central meaning and life of a purpose beyond its own end.

A similarly brittle, instrumental presentation of intimacy is observable in the present-day discourse around relations between the generations, where the value of 'parenting' is boiled down to the extent to which a parent can provide a child with the optimal resources, opportunities, and attitudes. In *Our Kids: The American Dream in Crisis*, the US sociologist Robert Putnam (2015) investigates the divide between the life chances available to the children of America's wealthy middle class and those of the very poor. For Putnam, the problem of America's stagnating social mobility can only be solved by parents adopting particular practices that enhance their children's social and cultural capital. While he recognises that '[e]ven ideal parenting cannot compensate for all the ill effects of poverty on children, and even incompetent parenting cannot nullify all the advantages conferred by parental affluence and education' (Putnam 2015, p. 144), he nonetheless sees parent training as the only way forward. What follows is a presentation of relations within the family, and between children and others in the community, as a process akin to corporate networking, where particular practices and behaviours are valued according to the extent to which they can build an individual child's access to status and resources.

In the domain of intergenerational relations, the language familiar from eugenics seems to be enjoying a revival. For example, towards the end of *Our Kids,* Putnam asks, rhetorically, whether we should 'delink sex from childbearing through more effective contraception', before arguing for a re-stigmatisation of unplanned pregnancy on the grounds that '[c]hanging the norm from childbearing by default to childbearing by design might have a big effect on the opportunity gap' (Putnam 2015, p. 267). Here, the problem of 'parenting' is addressed through discouraging particular kinds of people, from particular backgrounds and in particular circumstances, from having children.

On both sides of the Atlantic, a current popular policy idea resides in the promotion of 'early intervention'. This approach sees professional intervention into the lives of babies born to families considered problematic as a solution, not only to the problems those children might face but also to the problems of society as a whole, by helping 'families break the cycle of harmful social problems'. As the influential report *Early Intervention: Smart Investment, Massive Savings*, by Graham Allen, MP, put it:

> Early Intervention is an approach which offers our country a real opportunity to make lasting improvements in the lives of our children, to forestall many persistent social problems and end their transmission from one generation to the next, and to make long-term savings in public spending. (Allen 2011, p. xi)

The language used by the Allen Report—as well as the concept behind early intervention—is unapologetically eugenic and portrays a coldly instrumentalist view of relationships between the generations. For example:

> One of the key concepts used when we are talking about the problems of dysfunction is that cold business phrase—**stock and flow**. Remedial or late intervention policies address the **stock** of people already suffering from deep-rooted problems. Early Intervention seeks to block, reduce or filter the **flow** of new people (babies, children, young people) entering the stock. The current balance of policy is simply wrong. (Allen 2011, p. xv; emphasis in original)

Work by Macvarish (2014, 2016) has confirmed the centrality of biologised assumptions to early intervention policy, and indicated the disruptive effect this has on the intimacy and privacy of relations between parent and child. One effect of such 'brain claims' on thinking about social problems, she argues, is to 'consolidate a profoundly pessimistic view of children's potential. If the years 0–3, or even 0–2, are indeed the most important in a person's life, then there is no scope for the older individual to transform themselves or for society to help in the later amelioration of disadvantages'. Not only do brain claims 'shut down any discussion about different ways of raising children,' argues Macvarish—'they also promise to make parental love directly measurable in the behaviour of their offspring' (Macvarish 2014, p. 181).

CONCLUSION

Interest in the sociology of generations in recent years has disproportionately focused on the Baby Boomer generation: a generation that was born immediately after the Second World War and came of age in the tumultuous era known as 'The Sixties', and that, since its birth, has attracted swathes of academic studies, media commentary, and cultural iconography (Bristow 2015). The *Zeitgeist* of the 1960s, which this generation is seen to embody and express, centrally involved a reaction against traditional norms of sex, gender, and sexuality.

The Boomers, self-conscious in their youthful disregard for the 'old ways' and, rejecting the politics of the past in favour of a direct, egotistical orientation towards history—summed up in the belief that 'the personal is political'—symbolised the integration of all aspects of social and personal life into public discourse. This cultural turn built on a crumbling of the 'separate spheres' ideology that underpinned traditional notions of the (male) breadwinner and (female) housewife: the positioning of reproduction as something that should be chosen, planned for, and controlled (primarily, by women) and the growing adoption or oversight of many of the functions of the family by other institutions, including the school, the law, and social services.

As a result of these cultural and structural changes, it is now inconceivable that any generational narrative should be complete without an acknowledgement of women's agency. Yet for women, the hard-fought struggle to gain authorship of their own life story happened in a context where the scope of that life story became more narrowly cast. The rise of a beleaguered, risk-conscious selfhood, described in the previous chapter, has narrowed the terrain of history-making in the public realm. The possibilities of intimacy and fulfilment within the private realm, meanwhile, have become constrained by the increasing demands that reproduction – both in its biological and social sense – be managed in accordance with the political imperatives of today, couched in the rhetoric of safeguarding the younger, and 'future' generations.

NOTE

1. Some of the arguments in this section have been previously discussed in Bristow (2006).

REFERENCES

Allen, G. (2011). *Early intervention: Smart investment, massive savings*. The Second Independent Report to Her Majesty's Government. Available at: https://www.gov.uk/government/uploads/system/uploads/attachment_data/file/61012/earlyintervention-smartinvestment.pdf. Accessed 13 Feb 2015.

Arendt, H. (1998 [1958]). *The human condition*. Chicago: University of Chicago Press.

Badenhausen, R. (2003). Mourning through memoir: Trauma, testimony, and community in Vera Brittain's "Testament of Youth.". *Twentieth Century Literature, 49*(4), 421–448.

Beck, U., & Beck-Gernsheim, E. (1995). *The normal chaos of love*. Cambridge: Polity Press.

Beck-Gernsheim, E. (2002). *Reinventing the family: In search of new lifestyles*. Cambridge: Polity Press.

Berger, B., & Berger, P. (1983). *The war over the family: Capturing the middle ground*. New York: Anchor Press/Doubleday.

Bristow, J. (2006, March 28). Are we addicted to love?. *spiked*. Available at: http://www.spiked-online.com/newsite/article/271#.VnQjgUqLTIU. Accessed 18 Dec 2015.

Bristow, J. (2010, April 28). Turning parents into "partners of the state". *spiked*. Available at: www.spiked-online.com/newsite/article/8665. Accessed 9 Dec 2015.

Bristow, J. (2015). *Baby boomers and generational conflict*. Basingstoke: Palgrave Macmillan.

Brittain, V. (1984 [1933]). *Testament of youth*. London: Virago.

Chitty, C. (2013). The educational legacy of Francis Galton. *History of Education, 42*(3), 350–364.

Cohen, D. A. (1993). Private lives in public spaces: Marie Stopes, the mothers' clinics and the practice of contraception. *History Workshop, 35*, 95–116.

Cole, J., & Durham, D. (2007). *Generations and globalization: Youth, age and family in the new world economy*. Bloomington/Indianapolis: Indiana University Press.

Daly, M. (2010). Shifts in family policy in the UK under New Labour. *Journal of European Social Policy, 20*(5), 433–443.

Davis, K. (1940). The sociology of parent-youth conflict. *American Sociological Review, 5*(4), 523–535.

Falkingham, J. (1997). Who are the baby boomers? A demographic profile. In M. Evandrou (Ed.), Baby boomers: Ageing in the 21st century (pp. 15–40). London: Age Concern England.

Furedi, F. (2001). *Paranoid parenting: Why ignoring the experts may be best for your child*. London: Allen Lane.

GENDER AND THE INTIMATE POLITICS OF REPRODUCTION 111

Giddens, A. (1992). *The transformation of intimacy: Love, sexuality and eroticism in modern societies.* Cambridge: Polity Press.

Gillard, D. (2011). *Education in England: A brief history.* Available at: www.educationengland.org.uk/history. Accessed 9 Dec 2015.

Gillies, V. (2011). From function to competence: Engaging with the new politics of family. *Sociological Research Online, 16*(4), 11. Available at: http://www.socresonline.org.uk/16/4/11.html. Accessed 18 Dec 2015.

Intergenerational Foundation. (2015). Intergenerational fairness index 2015. Available at: http://www.if.org.uk/archives/6909/2015-intergenerational-fairness-index. Accessed 10 Dec 2015.

Kent, S. B. (1987). *Sex and suffrage in Britain, 1860–1914.* Princeton: Princeton University Press.

Lee, E., Bristow, J., Faircloth, C., & Macvarish, J. (2014). *Parenting culture studies.* Basingstoke: Palgrave Macmillan.

Lesthaeghe, R. (2010). The unfolding story of the second demographic transition. *Population and Development Review, 36*(2), 211–251.

Lewis, J. (1980). *The politics of motherhood: Child and maternal welfare in England, 1900–1939.* London: Croom Helm.

Lewis, J. (2011). Parenting programmes in England: Policy development and implementation issues, 2005–2010. *Journal of Social Welfare and Family Law, 33*(2), 107–121.

MacInnes, J., & Díaz, J. P. (2009). The reproductive revolution. *The Sociological Review, 57*(2), 262–284.

Macvarish, J. (2014). Babies' brains and parenting policy: The insensitive mother. In E. Lee, J. Bristow, C. Faircloth, & J. Macvarish (Eds.), *Parenting culture studies* (pp. 165–183). Basingstoke: Palgrave Macmillan.

Macvarish, J. (2016, in press). *Neuroparenting and the destruction of parental love.* Basingstoke: Palgrave Macmillan.

Marks, L. V. (2010). *Sexual chemistry: A history of the contraceptive pill.* New Haven: Yale University Press.

Moscucci, O. (2003). Holistic obstetrics: The origins of "natural childbirth" in Britain. *Postgraduate Medical Journal, 79*, 168–173.

Neushul, P. (1998). Marie C. Stopes and the popularization of birth control technology. *Technology and Culture, 39*(2), 245–272.

Pankhurst, E. (2015 [1914]). *Suffragette: My own story.* [e-book edition] London: Hesperus Press Ltd.

Planned Parenthood Federation of America. (2004, October). Opposition claims about Margaret Sanger. Fact Sheet. Available at: https://www.plannedparenthood.org/files/8013/9611/6937/Opposition_Claims_About_Margaret_Sanger.pdf. Accessed 18 Dec 2015.

Preston, S. H. (1984). Children and the elderly: Divergent paths for America's dependents. *Demography, 21*(4), 435–457.

Pugh, M. (2000). The march of the women: A revisionist account of the campaign for women's suffrage, 1866–1914. Oxford: Oxford University Press.

Putnam, R. D. (2015). *Our kids: The American dream in crisis.* New York: Simon and Schuster.

Sanger, M. (1921, October). The eugenic value of birth control propaganda. *Birth Control Review*, p. 5. Available at: https://www.nyu.edu/projects/sanger/webedition/app/documents/show.php?sangerDoc=238946.xml. Accessed 9 Dec 2015.

Schwarz, L. (2001). Vera Brittain's *Testament of Youth*: In consideration of the unentrenched voice. *a/b: Auto/Biography Studies, 16*(2), 237–255.

Walker, A. (1996). *The new generational contract: Intergenerational relations, old age and welfare.* London: UCL Press.

Ward, S., & Eden, C. (2009). *Key issues in education policy.* Thousand Oaks: Sage.

Wellings, K., & Johnson, A. M. (2013, November 30). Framing sexual health research: Adopting a broader perspective. *The Lancet 382*(9907), 1759–1762.

White, J. (2013). Thinking generations. *British Journal of Sociology, 64*(2), 216–247.

Whiteside, N. (2012). The liberal era and the growth of state welfare. In P. Alcock, M. May, & S. Wright (Eds.), *The student's companion to social policy* (4th ed.). Oxford: Wiley Blackwell.

Willetts, D. (2010). *The pinch: How the baby boomers took their children's future—And why they should give it back.* London: Atlantic Books.

CHAPTER 6

Conclusion

Abstract Talk of a crisis of generations in the present day both overstates and underestimates the problem. Most empirical research indicates that, in practice, generations continue to support and care for each other, and there is little overt intergenerational conflict. However, the wider ambivalence about knowledge, adulthood, and intimacy reveals a deep sense of uncertainty about the possibility of any history outside of the isolated self, which is being transmitted to the younger generation in both explicit and implicit ways.

Keywords Fathers and Sons • Me Decade • Tom Wolfe • Future • Youth

In formulating the problem of generations as it appeared in the 1920s, Mannheim drew on the insights of positivist and qualitative approaches to develop a method that would synthesise biology, biography, and history. When analysing the problem of generations in the present day, we can similarly take on board the insights of demographic and life course approaches, while recognising the need for a method that situates these insights in their wider social and historical context. This accounts for Mannheim's enduring influence on the study of generations and the wider field of the sociology of knowledge.

The aim of this essay has been to account for some of the developments in the sociology of knowledge, and the wider social forces framing the problem of generations, over the course of the twentieth century.

J. Bristow, *The Sociology of Generations*,
DOI 10.1057/978-1-137-60136-0_6

By adopting a schematic approach comparing the *Zeitgeist* of the interwar years, the 1960s, and the present day, we have attempted to highlight continuities and changes in the ways in which the relations within and between generations have been conceptualised in culture and framed by politics and policy.

In this regard, the suggestion that some new directions and challenges affect our understanding of the problem of generations in the present day does not imply that we are witnessing brand new problems, or areas of life that have not been adequately theorised. The challenges and opportunities presented by globalisation, women's equality, ageing, and other significant developments of our time have been compellingly analysed elsewhere, in terms of the continuities and changes that they represent. As we have noted, many empirical studies have challenged the assumption that these changes, in their own terms, have presented a threat to relations between the generations.

Rather, the argument put forward by this essay is that the new directions and challenges for the sociology of generations arise from a wider crisis of knowledge, which frames agency and consciousness in highly individualised, instrumental terms. In this respect, members of the 'older' or 'younger' generations are positioned as needing to find their sense of self, not in relation to the wider social forces and personal connections that give life its expansive quality, but in opposition to them.

'The husband and wife who sacrifice their own ambitions and their material assets in order to provide "a better future" for their children… the soldier who risks his life, or perhaps consciously sacrifices it, in battle… the man who devotes his life to some struggle for "his people" that cannot possibly be won in his lifetime… people (or most of them) who buy life insurance or leave wills… and, for that matter, most women upon becoming pregnant for the first time… are people who conceive of themselves, however unconsciously, as part of a great biological stream,' wrote the US writer Tom Wolfe in his 1976 essay excoriating 'The Me Decade'. He continued:

> Just as something of their ancestors lives on in them, so will something of them live on in their children… or in their people, their race, their community—for childless people, too, conduct their lives and try to arrange their postmortem affairs with concern for how the great stream is going to flow on. Most people, historically, have *not* lived their lives as if thinking, 'I have only one life to live.' Instead they have lived as if they are living their

ancestors' lives and their offspring's lives and perhaps their neighbors' lives as well. They have seen themselves as inseparable from the great tide of chromosomes of which they are created and which they pass on. The mere fact that you were only going to be here a short time and would be dead soon enough did not give you the license to try to climb out of the stream and change the natural order of things. (Wolfe 1976)

The sociologist of generations might object to the idea that the meaning of life lies in its biological reproduction—'the great tide of chromosomes'. We might also object to the Burkean presumption that the generational continuity can, and should, only be preserved if nobody tries to 'climb out of the stream and change the natural order of things'. As this essay has argued, it is precisely the friction between the generations, as representatives of their moment of history, which provides for the dynamism of knowledge and the continual renewal of our world.

But Wolfe is right to emphasise the fallacy at the heart of the beleaguered self—that 'I only have one life to live' and that this is the life of the present day. Indeed, it is the contact, and conflict, between the generations that gives individuals, who are 'only going to be here a short time and [will] be dead soon enough', the capacity to transcend our moment. We live in the present, but with one foot in the past and the other in the future.

The conflict of generations is rarely so stark as that dramatized by the Russian novelist Ivan Turgenev (1996 [1862]) in his classic novel *Fathers and Sons*, where both Bazarov, the young nihilist wedded to the virtues of science and progress, and Pavel Petrovitch, the conservative defender of aristocracy and the old ways, find themselves slain as a consequence of their beliefs. Turgenev situates the clash between past and present, between the old ways and the 'nihilists' (new men), within his central characters, depicting the ways in which such conflict is both exacerbated and tempered by its embodiment within living members of different generations.

Jahn (1977) describes Bazarov and Pavel Petrovich as examples of 'the proverbial similarity of opposites'—they are both as stubborn as each other. 'I'm convinced... that you and I are far more in the right than these young gentlemen, though we do perhaps express ourselves in old-fashioned language, *vieilli*, and have not the same insolent conceit,' argues Pavel Petrovich in *Fathers and Sons*:

Indeed, what a puffed-up crowd is the youth of today! Should you ask one of them whether he will take white wine or red, he will reply, in a bass voice, and with a face as though the whole universe were looking at him: "Red is my customary rule." (Turgenev 1996, p. 43)

Both these 'stubborn' representatives of their generation meet a similar fate, in a premature death. In this way, Turgenev expresses the importance of an open, constructive relationship between the generations, which stresses the need for accepted wisdom, transmitted from fathers to sons, but also the need for the older generation to accept that the younger generation will take that wisdom and mould it in its own way.

When generations are distanced from each other, through the bureaucratic mechanisms and cultural dynamics described in this book, such open and constructive relationships cannot flourish. The 'insolent conceit' of the younger generation is pre-empted by a culture that already imbues received wisdom with suspicion and promotes a vision of the future constrained by trepidation. The message relentlessly communicated by this presentist sensibility is, 'I have only one life to live'—a message that simultaneously discourages the younger generation from renewing their world, and from changing it.

References

Jahn, G. R. (1977). Character and theme in 'Fathers and sons'. College Literature, 4(1), 80–91.
Turgenev, I. S. (1996). *Fathers and sons*. Hertfordshire: Wordsworth Editions.
Wolfe, T. (1976, August 23). The 'Me' Decade and the Third Great Awakening. *New York Magazine*. Available at: http://nymag.com/news/features/45938/. Accessed 18 Dec 2015.

REFERENCES

Giroux, H. A. (2003). *The abandoned generation: Democracy beyond the culture of fear*. Basingstoke: Palgrave Macmillan.

Newbolt, H. (1892). Vitaï Lampada. Available at: http://www.poemhunter.com/poem/vita-lampada/. Accessed 5 Jan 2015.

Marwick, A. (1970). Youth in Britain, 1920–1960: Detachment and commitment. *Journal of Contemporary History 5*(1) Generations in Conflict: 37–51.

Marwick, A. (1999). *The sixties: Cultural revolution in Britain, France, Italy and the United States, c.1958–c.1974*. Oxford: Oxford University Press.

Thomas, N. (2002). Challenging myths of the 1960s: The case of student protest in Britain. *Twentieth Century British History, 13*(3), 277–297.

Gard, M., & Wright, J. (2005). *The obesity epidemic: Science, morality and ideology*. London/New York: Routledge.

Somerville, J. (2000). *Feminism and the family: Politics and society in the UK and USA*. Hampshire/London: Macmillan Press.

Brittain, V. (1984 [1933]). *Testament of youth*. London: Virago.

Falkingham, J. (1997). Who are the baby boomers? A demographic profile. In M. Evandrou (Ed.), *Baby boomers: Ageing in the 21st century* (pp. 15–40). London: Age Concern England.

Pugh, M. (2000). *The march of the women: A revisionist account of the campaign for women's suffrage, 1866–1914*. Oxford: Oxford University Press.

Jahn, G. R. (1977). Character and theme in 'Fathers and sons'. *College Literature, 4*(1), 80–91.

© The Editor(s) (if applicable) and The Author(s) 2016
J. Bristow, *The Sociology of Generations*,
DOI 10.1057/978-1-137-60136-0

INDEX

© The Editor(s) (if applicable) and The Author(s) 2016
J. Bristow, *The Sociology of Generations*,
DOI 10.1057/978-1-137-60136-0

Printed by Printforce, the Netherlands